Soil and Water Analysis in Environmental Sciences

THE AUTHORS

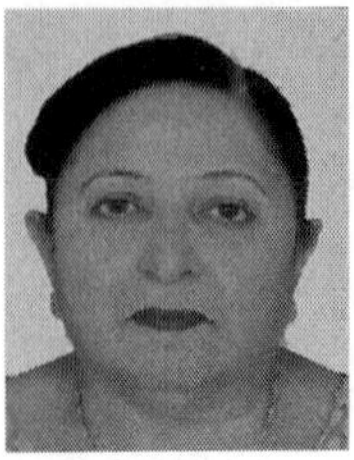

Dr. (Mrs.) Neelam Saharan was born on 6th June, 1957 in Karnal district of Haryana state. After completing schooling and graduation in Karnal, she did her post-graduation in Organic Chemistry and doctorate in Dairy Chemistry from Kurukshetra University, Kurukshetra, Haryana. She joined as scientist in 1983 at ICAR-Indian Aquaculture Research Institute, New Delhi and worked at ICAR-Central Research Institute for Dryland Agriculture and ICAR-Central Institute of Fisheries Education indifferent capacities and retired as Principal Scientist in 2019 from ICAR-CIFE, Mumbai. She has worked in 20 institutional and 6 externally funded research projects on soil and water management, nutrient management, reservoir development, pen and cage aquaculture, tribal development and bio-remediation of pesticides. She has written 130 research papers in various national and international journals, 20 review articles, 6 books, 8 training manuals and has received 7 awards for her distinguished work in water pollution and its control and soil and water quality management and amelioration of pesticides; institution building etc. She has guided 25 M.F.Sc. students and 12 Ph.D. students as major advisor. Presently she is working as professor emeritus at ICAR-Central Institute of Fisheries Education, Mumbai.

Dr. (Mrs) Vidya Shree Bharti was born on 10th July 1977 in the Madhubani District of Bihar. She completed her schooling from Jawahar Navodaya Vidyalaya, Ranti, Madhubani, BSc(Ag) Hons from Bidhan Chandra Krishi Viswavidyalaya, Nadia, W. B, MSc (Soil Science) from G.B. Pant University of Science and Technology, Uttrakhand and Ph.D. from Indian Institute of Technology, Mumbai, India. She joined her Agricultural Research Service as a Scientist in the Indian Council of Agricultural Research on 27th June 2005. She served as a Scientist for two years at the Indian Institute of Soil Science, Bhopal, and on 24th December 2007 she got transferred to the Central Institute of Fisheries Education, Mumbai. She published articles, book chapters, training manuals, and the papers in National and International journals. Since 2012 She is working as a senior scientist in the Aquatic Environment and Health Management division of ICAR- CIFE Mumbai and focusing on her research in the area of soil-water management for aquaculture,biochar application in aquaculture, climate change study mainly carbon sequestration and greenhouse gas estimation and mitigation in aquaculture, pollutant remediation with biochar, remote sensing and GIS application for Aquatic Environment management and precision Aquaculture. She has guided 13 MFSc students as a major guide and more than 20 as a member committee of MFSc and Ph.D. She visited CSIRO environmental lab, Australia as visiting scientist in 2022. She had worked on 11 projects and was awarded Geomatrix 2012, the INRIA Best paper award at IIT, Mumbai in 2012, and the Japan society of photogrammetry and Remote sensing outstanding Paper Award in Bali Indonesia in 2013.

Soil and Water Analysis in Environmental Sciences

Neelam Saharan

Vidya Shree Bharti

2024

Daya Publishing House®

A Division of

Astral International Pvt. Ltd.

New Delhi – 110 002

ISBN: 978-93-5461-990-8

Published by : **Daya Publishing House®**
A Division of
Astral International Pvt. Ltd.
– ISO 9001:2015 Certified Company –
4736/23, Ansari Road, Darya Ganj,
New Delhi-110 002
Ph. 011-43549197, 8130496929
E-mail: info@astralint.com
Website: www.astralint.com

Printed at : **Replika Press Pvt. Ltd.**

Foreword

Soil is a non-renewable resource upon which mankind depends for survival. Soil is a key factor in aquaculture. Most of the pond is built from and in soil. Many dissolved and suspended substances are derived from contact with soil. Pond soil is a storehouse for many substances that accumulate in the pond ecosystem and chemical and biological process occurring in the surface layer of pond soil influences water quality and aquaculture. Hence an understanding of soil properties and the reactions and processes in the soil can be useful in pond aquaculture. The management of soil and water quality is an important factor for the sustenance of our food security.

The soil and water analysis book will provide comprehensive techniques, updated methods, and standards for sample collection, handling, and analysis at the site and in the laboratory for practical and research purposes. It provides information on how to collect sediment and water samples for analysis of different quality parameters that can be measured in the field and transportation of samples to the laboratories for further analysis. To be meaningful and valid, soil and water tests have to be appropriate for the purpose intended and reliable and repeatable. It is considered an important tool to assess nutrient needs and formulate specific nutrient recommendations.

The book prepared for the soil and water analysis is a compilation of lectures and important relevant protocols. I am sure that this would serve as useful reference material for students, state fisheries officers, progressive farmers, entrepreneurs, and laboratory technicians who are interested in soil and water quality analysis in aquaculture and fisheries. I congratulate the authors, Dr. Neelam Saharan and Dr. Vidya Shree Bharti, for their endeavor in excellently bringing out this publication.

ICAR-CIFE, Mumbai

Ravi Shankar C. N.
Director and Vice Chancellor
ICAR-CIFE, Mumbai

Preface

A comprehensive manual on Analytical Chemistry to cater to the needs of Fisheries students is a long-felt need as the literature dealing with soil and water analysis is scanty and scattered. Analytical chemistry in the recent past has been dominated by instrumental analysis and the use of conventional methods of analysis will be rare in the future. The present book contains most of the standard analytical methods that are used at the Department of Aquatic Environment Management (AEM) of the Central Institute of Fisheries Education, Mumbai. In this book, we have striven for a more complete coverage of the various analytical methods and pre-treatments used in our Department. Therefore, this book contains, in addition to procedures for physico - chemical soil and water analysis, also procedures for CHNS analysis, Biochar production, and characterization, C:N ratio determination, and Bio-synthesis of the Nano-particles. As many of the texts in this book are new, they can be improved.

We have highly sophisticated tools like inductivity coupled plasma–atomic emission spectrophotometry, atomic absorption spectrophotometry, auto analyzer, etc. where several elements can be analyzed within a short span of time with great accuracy. However, in many of the analytical laboratories imparting environmental education, such sophisticated techniques cannot be implemented in the developing world. Therefore, one has to depend on conventional methods of analysis which can be followed with great ease and accuracy. The analysis of soil and water samples is done with a view to characterize them qualitatively and quantitatively.

With this view in mind, this manual has been compiled describing various methods of analysis for soil and water analysis besides presenting the basic principles of analytical chemistry. Of course, one cannot be original in describing various methods of analysis. It can only be a compilation in one's own way. The calculation part is explained step by step so that the students can very well understand the basics involved. The book consists of two sections covering methods of soil and water analysis.

The authors are indebted to the Indian Council of Agriculture Research, New Delhi for providing the funds from Emeritus Professor Scheme to publish this book. The authors owe a great deal to the guidance and encouragement received from Dr. C.N. Ravishankar, Director, ICAR-CIFE, and Dr. K. K. Krishnani, Head, AEHM at ICAR-CIFE, Mumbai. The authors profusely thank Dr. Dasari Bhoomaiah for designing the cover page of the book. It is hoped that this book will serve the need to fulfill a self-contained laboratory manual in soil and water testing methodology and will provide the reader with not only a detailed description of the analytical procedures but also insight into understating aquatic analytical chemistry. The authors would the most grateful for any comments, corrections, and additions.

ICAR-CIFE, Mumbai

Dr. Neelam Saharan

Dr. Vidya Shree Bharti

Contents

Introduction

The management of soil and water quality is the singular most important factor in productive aquaculture. It is defined as the network of chemical, biological, and physical factors that constitute the soil and water environment and influence aquaculture practices. Hundreds of thousands of chemical substances can be found in association with soil and water. They range from such common ions as sodium and chloride which account for much of the salinity in the sea to trace elements such as cadmium, copper and gold and include various naturally occurring and manufactured organic compounds like CO_2, H_2S, and iron at levels that make the soil and water unsuitable for use unless the undesirable chemical is reduced to a safe level or completely removed. The effects of metals in soil and water range from beneficial through troublesome to dangerously toxic. Some metals are essential while others may adversely affect water consumers. Some metals may be either beneficial or toxic depending upon their concentration. Soil and water quality criteria specify the concentration of the constituents which, if not exceeded, are expected to result in an aquatic ecosystem suitable for productivity process and growth of organisms under culture. Such criteria are derived from experimental or in situ observations that depict organism responses to a defined stimulus or material under identifiable or regulated environmental conditions for a specified time period.

In aquaculture, soil, and water quality is determined to know their total nutrient load and their suitability for the growth and survival of fish and is usually governed by several variables. Soil and water analysis is carried out for useful as well as harmful constituents. The latter is particularly important for determining a value for fish survival, health, and environmental angles. The standard methods commonly used for determining the total salts, pH, cations, and anions that are important for evaluating soil and water quality, salinity hazard, fish nutrition and environmental points of view are discussed here. The limnochemist's working conditions vary from solitary fieldwork (perhaps in a remote area) to a well-equipped laboratory with abundant expertise at hand. Therefore, one or more methods are described in this manual. These correspond to the differences in the available equipment and exactitude though different methods are often variations on the same theme. The success of soil and water testing depends to a great extent on the methodology adopted and the interpretation of test results. A suitable method is one that satisfies the three-tiered criteria of accuracy or reproducibility, rapidity or speed, and predictability or relationship of test values with aquatic production (Singh et al., 1999). For successful aquaculture of fish, molluscs, crustaceans, and even aquatic plants, the soil and water in which they are grown, should offer favourable conditions for their growth, which in turn, are intimately linked with their physical, chemical, and biological characteristics.

Chapter 1

SOIL TESTING

Soil constitutes the solid surface layer of the earth upon which humans conduct most of their activities and everyone has some familiarity with the features of the land surface and the soil. Pond soils have various roles in the production of fish, the most important one being their ability to store nutrients and release them into water under different reaction mechanisms. Soils also have a paramount role in maintenance of favorable chemical conditions viz. pH, alkalinity, hardness, electrical conductivity, dissolved gases, etc., and in controlling the adsorption–desorption and bio-availability of heavy metals and toxic elements and hence are rightly described as 'Chemical Laboratory of Ponds'. Therefore, analysis of pond soil and maintenance of soil quality are essential while developing strategies for pond productivity and water quality. Some selected parameters such as determination of soil texture, soil reaction, electrical conductivity, organic carbon, exchangeable potassium, total and available nitrogen, phosphorus, and available micronutrients viz. Zn, Cu, Fe, Mn, B, and Mo are usually analyzed to evaluate soil properties that are known to have a direct impact on fish production.

Soil Sampling

The collection of representative and composite samples of soil is as important as the analysis of the sample itself. Hence considerable care must be devoted to the selection and collection of samples since they form the basis of subsequent laboratory descriptions of the soil. As a very small fraction of the huge soil mass is used for analysis, it is essential to get a truly representative soil sample. For making a composite sample, collect small portions of soil up to the desired depth with a suitable sampling tool from 8-20 well-distributed spots depending upon the area, moving in a zig-zag manner (Jones, 1988). For the collection of soils from lake bottoms, Van Veen grab sampler and Ekman's dredge for soft soils and Peterson's grab for hard earth are generally employed. Mix the soil collected from various spots covering the entire area thoroughly on a polythene sheet. Reduce bulk to 500 g by the quartering process in which the entire soil mass is spread, divided into four quarters, two opposite ones are discarded and the remaining two are remixed. Repeat this process until about 500g of soil is left. The sample must be kept in a polythene bag with suitable description and identification marks.

Size of Sample Material

The common size for the sample material is the fine earth fraction (soil fraction < 2 mm), except when mentioned otherwise in the text. When very small sample quantities (<100 mg) should be used, the sample should be ground.

Chemicals

All chemicals used in the procedures should be of analytical grade (p.a.) quality except when otherwise mentioned in the text. The water may be distilled water except when the text indicates ultra-high quality.

Apparatus

Under the headlines "apparatus" in the procedures, only special equipment is mentioned. Standard laboratory equipment, such as glassware and tubes is supposed to be present in any laboratory.

Quality

Every series for analysis should consist of the samples, two blanks and 1 or preferably 2 reference samples to check the performance of the analysis. When the results of these reference samples are beyond the tolerated deviation, the analysis should be repeated. Another way to check the performance of a laboratory is to join an inter-laboratory sample exchange program.

Why Soil Testing?

- ☆ Soil fertility status assessment involves the estimation of its available nutrient status
- ☆ It gives the amount of nutrients directly available in the soil for subsequent uptake by fish.
- ☆ Guides to arrive at optimum fertilizer application.
- ☆ It is a method of evaluating the nutrient status (physico-chemical properties) of the soil i.e assessment of the fertility of the soil to determine nutrient deficiencies.
- ☆ It is also concerned with environmental quality for the community hazards.

Soil fertility may be defined as the capacity of soil to furnish available nutrients to the fish in the proper amount and appropriate balance, under the ideal conditions of plant growth whereas, soil productivity is the capacity of soil to produce under specific condition of crop production.

Advantages of Soil Testing

- ☆ More rapid method as compared to biological or deficiency symptoms/ plant analysis
- ☆ One may determine the need of the soil before aquaculture activity.
- ☆ To determine the suitability of the soil for laying gardens.
- ☆ Lime problems.
- ☆ Soil survey.

The error in soil sampling in a field is generally greater than the error in laboratory analysis. The most recommendation call for soil testing of each field is about every three years with more frequent testing on lighter soils. Therefore, it is necessary that the soil sample should be representative of the area. Further, the subsequent handling operation in the laboratory should be carefully performed because a minute quantity (1 – 10g) of the large soil mass of the field is actually used for the analysis in the laboratory. Unless one is sure of representative and proper sampling, the results obtained in the laboratory analysis will be of no use under the field conditions.

Precautions

1. Do not draw any sample from the extreme corners of the field, areas recently manured or fertilized, old bounds, and marshy spots.
2. Avoid sampling from furrows, and acidic or alkaline pockets.
3. Keep the sample in a bag and tag it properly.
4. Do not take less than 0.5 kg of a composite sample.
5. Sampling should be done from a uniform piece of land.
6. If there is a hard pan in the pit, it should be sampled separately, and also notedown its depth and thickness.

Van Veen grab when it is locked, ready to be lowered to the floor

It is now unlocked after hitting the floor, ready to collect sediment

When lifting it after unlocking, the jaws will close to collect sediment inside

Apparatus and materials for soil sampling:

<table>
<tr><th>Apparatus</th><th></th><th>Material</th></tr>
<tr><td>Khurpi</td><td>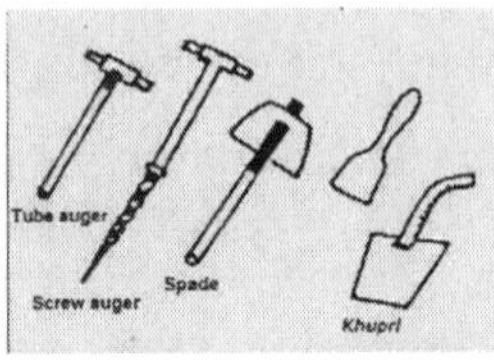
</td><td>Polythene/paper/cloth bags
Scale
Plastic bowl
Mortar and pestle
Cardboard cartons
Labels
Wooden roller</td></tr>
<tr><td>Spade</td><td>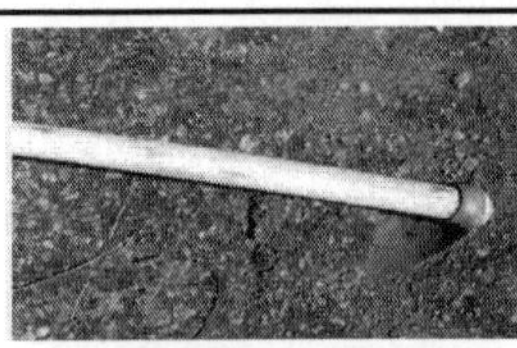</td><td></td></tr>
<tr><td>Auger</td><td>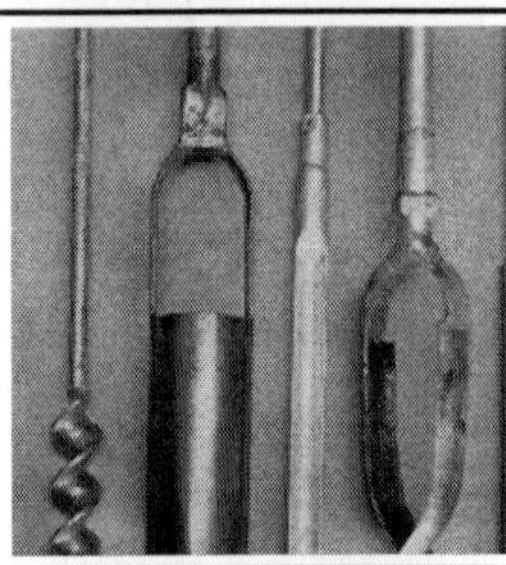</td><td></td></tr>
<tr><td></td><td>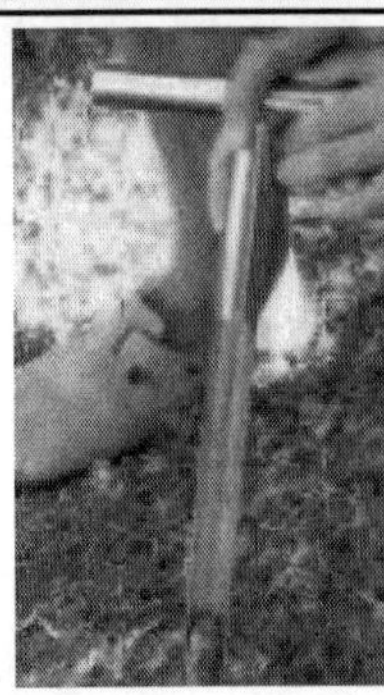</td><td>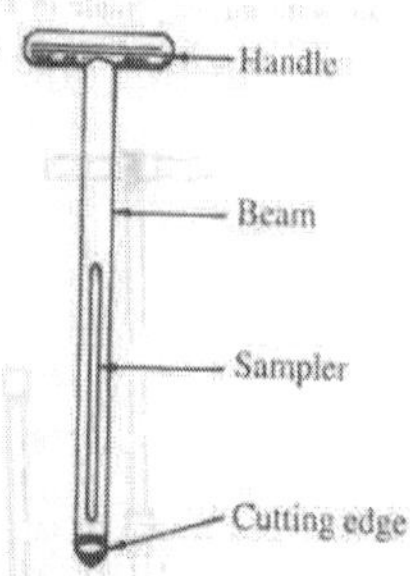

Figure 2.2 : Parts of an auger</td></tr>
<tr><td>Scale for measuring 6' depth</td><td>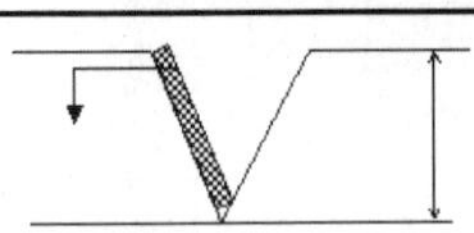
1 inch / 2.5 cm 6 inches (15 cm)</td><td>
Sieve</td></tr>
</table>

1.1 Soil Processing

When the soil sample reaches the laboratory, break any large lumps and spread it out in a large aluminum tray to become air dry. If desired, keep one or two of the larger lumps to illustrate structure or colour. It is preferable to air dry the soil samples under a shed at 20-25 °C. Samples may be mixed during drying to expose fresh surfaces. After air-drying, the soil samples are gently crushed in a pestle and mortar and sieved through a 2 mm sieve. Crushing should be continued until the soil retained on the sieve contains no aggregates. Remix the entire quantity of sieved soil thoroughly and store it in a suitable sample bottle. For identification, place a label on the bottle as well as inside.

1.2 Moisture Content

Principle

Since the results of soil analysis are to be calculated on the basis of oven-dry sample weight, it is important to execute moisture analysis before any other analysis. The result on the basis of the air-dry weight is multiplied by a moisture correction factor (mcf). This recalculation is not valid for analyses with a measurement obtained directly in the extract (e.g. Fe and Al fractionation and KCl extractable ammonium and nitrate). If the additional amount of soil moisture is neglected, there will be a difference in the final result. So when an analysis contains a direct measurement in the extract, the additional soil moisture has to be included in the calculation (Burman et. al., 1996).

Apparatus

Drying oven, porcelain crucible, and electronic balance.

Procedure

Place a crucible in the oven at 105 °C and leave it for at least two hours. Then place it in a desiccator and let it cool down to room temperature. Weigh the empty crucible (weight A). Weigh at least 2g of sample in the crucible and weigh it again (weight B). Place the crucible overnight in the oven at 105 °C. Transfer the crucible to the desiccator and allow to cool to room temperature. Finally, weigh the crucible again (weight C).

Calculation

$$\text{M (Moisture Content) \%} = \frac{B - C \times 100}{C - A}$$

$$\text{M C F (Moisture Correction Factor)} = \frac{100 + M(100)}{100}$$

1.3 Bulk Density of Given Soil

Requirements - Oven, measuring cylinder, and chemical balance.

Method- Dry the soil sample in the oven at 105^0C until a constant weight is attained, transfer a little dried soil to a measuring cylinder, and note the volume. Record the weight of this volume of soil on a balance.

Observations

S.No	Weight of soil (gm)	Volume of soil (ml)	Bulk density (gm/cm3)
1			
2			
3			
4			

Calculation

Bulk density (gm/cm^3) = Weight of soil (gm)/Volume of soil (cm^3)

Result- The bulk density of the given soil was observed ------- gm/cm^3.

Precautions

1. Note the value carefully.
2. Weigh the soil exactly.
3. Dry the soil properly

1.4 Water Holding Capacity of a Given Soil

Requirements- Oven, perforated circular soil boxes (5.6 cm diameter, 1.6 cm height, bottom perforated with holes of 0.75 mm diameter), Filter paper (Whatman no.1), Petridish and electronic balance.

Method

- ☆ Dry the crushed soil sample in an oven at 105^{0}C.
- ☆ Place a filter paper inside the perforated bottom of the circular soil box, weigh the box, and fill it with a dried soil sample.
- ☆ Note the weight of the box filled with dried soil.
- ☆ Place the box in a petri dish of 10 cm diameter containing water for about 12 hours so that the water enters the box and saturates the soil. Take the box out of the water wipe it dry on the outside and record its weight.

Observation

S. No	Initial weight of soil box (gm) W^0	Weight of soil box with soil (gm) W^1	Weight of soil box with soil after absorbing water (gm) W^2	Calculation of Water holding capacity (%)
1				
2				
3				
4				

Water holding capacity (%) = (W2-W1)-(W1-W0)/(W1-W0)*100

Result- The water holding capacity of given soil was observed -----------%.

Precautions

1. Weight the soil and write the particulars properly.
2. Dry the soil box from outside properly.

1.5 Soil Texture

The mechanical analysis of soil gives an idea of the texture of the soil. It consists essentially of two operations viz. dispersion of soil and grading the dispersed particles into size groups. The texture of soil refers to the distribution of the different sizes of soil particles. Soil texture is qualitatively determined by a rapid feel method that involves rubbing the soil between the thumb and forefinger (Singh et al., 1999). In this procedure, proficiency is gained through practice and making a comparison with samples of known textural classes determined by quantitative method viz. Bouyoucos hydrometer method (Bouyoucos, 1962).

Rapid Feel Method

A small quantity of air-dried soil is moistened with water and mixed thoroughly on a glass or porcelain dish to form a soft ball and then worked until stiff and squeezed out between thumb and forefinger. The feel to finger, ease of forming a ball, stickiness or grittiness, whether forming soil ribbon or merely crumbling on squeezing, etc. are observed. The sample is assigned approximate textural class as given below:

I. Very Coarse Texture

Sandy: Very gritty, does not form ball, does not stain finger.

II. Coarse Texture

Loamy Sand: Very gritty, forms ball but very easily broken, stains fingers slightly.

III. Medium Texture

Loam: Neither very gritty nor very smooth, forms firm ball but does not ribbon, stains finger appreciably.

Silt loam: Smooth and sticky, buttery feel, forms firm ball, stains and has a slight tendency to ribbon with flaky surface.

IV. Fine Texture

Clay loam: Moderately sticky, slightly gritty feel, forms moderately hard ball when dry, stains, ribbons out on squeezing but the ribbon breaks easily.

Silty clay loam: Same as above but very smooth, shows flaking on ribbon surface, similar to silt loam.

Clay: Very sticky feel, forms a ball which when dry cannot be crushed by fingers, stains heavily, squeezes out at right moisture into long (2.5 cm) ribbon.

Presently, the best description of texture that can be derived from the mechanical analysis, is that based on the triangular diagram of Prescot, Taylor and Marshall (1934) shown in Figure. The percentage distribution of sand, silt and clay is determined according to Bouyoucos Hydrometer method (Bouyoucos, 1962) and then fits into the triangular diagram given below to ascertain the textural class of the given soil sample.

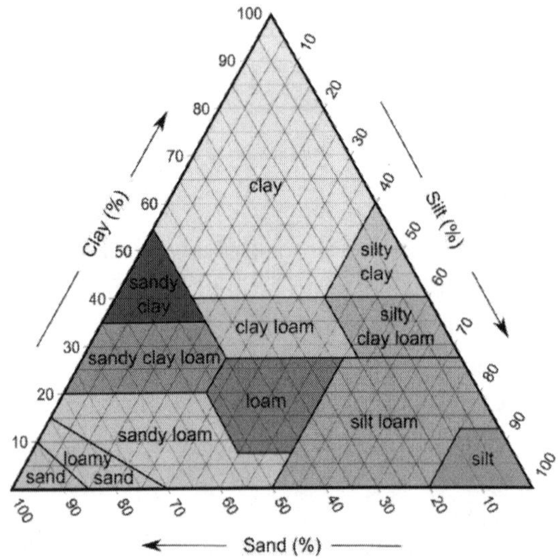

Textural Triangle

☆ The horizontal lines mark percentages of clay (by mass). The lines angle upwards to the right mark percentages of silt. The lines angled upwards to the left mark percentages of sand.

☆ The percentage of silt and clay are used to draw lines. When three lines so drawn intersect in a compartment then it is the textural class of the soil under study.

Bouyoucos Hydrometer Method

Principle

The method is based on 'Stokes Law' which states that rate of fall of particles in a suspension is directly proportional to their size. Accordingly, larger particles will settle first followed by silt and then by clay. The resistance offered by the liquid to the fall of particles varies with the radius of the sphere. The law is expressed as:

$$V = \frac{2gr2(D\text{-}d)}{9n}$$

Where	V	=	Velocity of the fall of particle (cm sec^{-1}).
	g	=	acceleration due to gravity.
	d	=	density of the particle.
	D	=	density of the liquid and
	N	=	absolute viscosity of the liquid.

Assumptions of Stoke's law:

☆ The particles must be large enough, so that Brownian movement will not influence their rate of fall.

☆ There must be no slipping between the liquid and the particles.

☆ The velocity of fall must not exceed a certain critical value.

☆ The particles fall independently. This requires a suspension concentration of less than 5 per cent.

☆ The particles must be rigid and smooth.

Limitations of Stoke's law:

1. It is necessary to maintain a constant temperature during analysis in the vessel filled with suspension. Otherwise, there will not be uniform settling of the particles and results will be wrong. This is not possible most of the times.
2. Since the density of clay particles decreases with the particle size, hence it may affect the accuracy of the Stoke's law.

Apparatus

Hydrometer, mechanical stirrer, thermometer, suspension cylinder with 34 + 2 cm height and up to 1000 ml mark.

Reagent

10% Sodium hexametaphosphate.

Procedure

Transfer 100 g of a soil sample to a 600 ml conical flask. Add 200 ml of distilled water, and 100 ml of the sodium hexametaphosphate solution, stir well with a glass rod, cover, and keep for 4 to 5 h preferably overnight. Transfer the contents quantitatively to the cup of the mechanical stirrer giving 4-5 washings of distilled water to transfer all soil particles, making the volume to about 500 ml. Run the stirrer for ten minutes and transfer the contents to the suspension cylinder (1000 ml) giving 4-5 washings of distilled water and making the volume up to mark. Put the rubber stopper tightly and invert the cylinder carefully. Vigorously shake several times to allow the soil particles to disperse thoroughly. Keep the cylinder on the table. Remove the stopper. Immediately place the hydrometer in the suspension, gently checking any up and down movement. Record the reading exactly 40 seconds after placement of the hydrometer. Replace the rubber stopper and invert several times again to ensure complete dispersal of particles. Keep it on the table. Exactly after 2 hours again place the hydrometer in the suspension and note the hydrometer reading. Simultaneously, run a blank without soil and record the room temperature in °F. In case the surface of the suspension in the cylinder is not clear due to foam, add 2-3 drops of amyl or capryl alcohol.

Calculation

Correction Faction (CF) = (Actual room temp. in °F-68) x 0.2.

$$\text{Percent Silt +Clay} = \frac{(S\text{-}B) + C\,F}{\text{wt. of sample (g)}} \times 100$$

where,

S and B stand for sample and blank readings respectively, taken after 40 second (first reading).

$$\text{Percent Clay} = \frac{(S\text{-}B) + C\,F}{\text{wt. of sample (g)}} \times 100$$

where,

S and B stand for samples and blank readings, respectively, taken after 2 hours

Percent sand = 100 - (Silt + Clay).

International Pipette Method of Texture Determination (Piper, 1966)

The pipette method is a standard method for particle size analysis of soils because of its accuracy, but it is time-consuming and cannot be employed where large numbers of samples have to be analysed. Particle size analysis is done by using sieves to separate out coarse sand from the finer particles. The silt and clay contents are determined by measuring the rate of settling of these two separates from the suspension in water. Using a pipette, samples are drawn at a given depth (10 cm) after a specific time and then the dry matter is determined.

Principal: The analytical procedure by which the proportion of soil particles of different sizes is determined is called mechanical analysis. Before mechanical analysis, the complete dispersion of soil into ultimate particles is necessary. Then the coarse sand is separated by sieving, the silt and clay by their rates of settling in a viscous medium, and the fine sand by decantation. This method is based on the principle that larger particles fall faster through water than smaller particles. This principle is quantified in Stoke's Law.

The sample preparation consists of removing materials that are not to be considered in the analysis. Particles larger than 2 mm in diameter are removed by sieving. Precise results require that the organic matter present is removed, usually by oxidizing it with hydrogen peroxide. Organic matter binds the clay particles together causing them to be measured as groups. Calcium carbonate in the soil which has a similar effect may be removed by treating the samples with dilute acids.

Apparatus

- ☆ The pipette sampling apparatus.
- ☆ An electric stirrer with a bowl.
- ☆ Sieves
- ☆ 600 ml beaker
- ☆ One litre – marked sedimentation cylinder.
- ☆ A wide – mouthed glass funnel for holding the sieve.
- ☆ Weighing beakers or weighing dishes
- ☆ A hot water-bath or a hot-plate.

Sedimentation times for particles of silt and clay through water for the 10 cm depth

Temperature (°C)	Settling time			
	Clay particles (2µm)		Silt particles (20 µm)	
	Hours	Minutes	Hours	Minutes
10	10	25	6	20
11	10	10	6	20
12	9	50	6	0
13	9	35	5	50
14	9	20	5	40
15	9	5	5	30
16	8	50	5	20
17	8	35	5	10
18	8	25	5	0
19	8	0	5	0
20	8	0	4	48
21	7	50	4	40
22	7	20	4	30
23	7	25	4	30

Temperature (°C)	Settling time			
	Clay particles (2μm)		Silt particles (20 μm)	
	Hours	Minutes	Hours	Minutes
24	7	15	4	20
25	7	15	4	15
26	6	55	4	10
27	6	45	4	5
28	6	40	4	0
29	6	30	3	55
30	6	20	3	50

Reagents

30% hydrogen peroxide

2 NHCl

N/10 NaOH

Procedure

1. Transfer 20 g of oven-dried soil to a 600 ml beaker.
2. Add 20 ml of H_2O_2 and swirl the contents well. Let the reaction proceed for 5-10 minutes.
3. Place the beaker over a hot water-bath or a hot plate. Continue digestion, stirring the contents all the time with a glass rod to minimize frothing, till the reaction completely subsides. In the case of severe frothing and prolonged reaction, a second lot of H_2O_2 should be added. Generally, the pond sediments contain high amounts of organic matter, a second treatment with H_2O_2 is usually required.
4. Cool the beaker; wash its inner sides with distilled water.
5. Add 25 ml of 2 N HCl and allow the contents to react. If more than 2% $CaCO_3$ is present, more HCl should be added at the rate of 2.5 ml for each percent of $CaCO_3$. Ensure that HCl reaction with the soil is complete. Therefore, allow the reaction to proceed for about an hour, with intermittent shaking at intervals.
6. Filter the contents through Whatman No. 50 filter paper. Discard the filtrate.
7. Give washings to the soil with distilled water till the filtrate coming out is free from chlorides. The test may be made by using a silver nitrate solution.
8. Transfer the soil sample from the filter paper to the bowl of the electric

stirrer with a jet of distilled water. Make the volume about 400-500 ml. Add 5-6 drops of phenolphthalein indicator. Add N/10 NaOH till the whole suspension shows a pink colour, indicating its alkaline reaction. Stir the contents for 10 minutes.

9. Transfer the contents to a 70-mesh sieve placed in the funnel, which is held over the top of a one-litre cylinder. Wash the soil particles on the sieve with a jet of distilled water till only the coarser particles are retained on the sieve.
10. Transfer these particles to a weighed dish. Dry at 105^{0}C in the oven to a constant weight and record this weight.
11. Make up the volume of the suspension in the cylinder to the one litre mark by adding more distilled water.
12. Place the cylinder upright on the table. Note down the temperature of the suspension.
13. Stir the contents 20-25 times in one minute with the plunger so that nothing remains settled at the bottom of the cylinder. Remove the plunger gently.
14. Find out the requisite time against the temperature for silt + clay and clay from the sedimentation table.
15. Pipette out a 25 ml sample of the suspension from 10 cm depth after the requisite time. Transfer the soil-water suspension to a weighed dish or beaker. Dry at 105^{0}C to a constant weight and record the weight.
16. In the same way as given above, remove 25 ml of the suspension from the 10 cm depth at the time required to sample for clay. Dry it in the oven.
17. Record this weight as the weight of the dish and the amount of clay in 25 ml of the suspension.
18. Decant the remaining suspension by washing repeatedly with distilled water.
19. Remove the particles finally left, with a jet of distilled water to the weighing-dish.
20. Dry the weighing dish with its contents in the oven. The weight of the soil particles in the dish is the amount of fine sand in the total sample.
21. Compare the weight of fine sand thus obtained with the weight calculated by subtracting the weight of coarse sand, slit, and clay from the total weight of the sample taken.

Observations

W_1	=	Weight of the container (beaker or weighing-dish used for coarse sand).
W_2	=	Weight of the container + the weight of coarse sand.
W_3	=	Weight of the container used for silt + clay.
W_4	=	Weight of the container + the weight of silt + clay
W_5	=	Weight of the container used for clay
W_6	=	Weight of container + the weight of clay
W_7	=	Weight of the container used for fine sand
W_8	=	Weight of the container + the weight of the fine sand

Calculations: - The four fractions of soil obtained above are finally calculated as below:-

$$\text{Percentage coarse sand} = \frac{W_2 - W_1}{20} \times 100$$

$$\text{Percentage silt + clay} = \frac{W_4 - W_3}{20} \times \frac{1000}{25} \times 100$$

$$\text{(c) Percentage clay} = \frac{W_6 - W_5}{20} \times \frac{1000}{25} \times 100$$

$$\text{Percentage fine sand} = \frac{W_8 - W_7}{20} \times 100$$

Percentage silt = b - c

Percentage fine sand by subtraction = 100- (a+b)

Precautions

- ☆ The temperature of the suspension between two pipettings should not be allowed to vary too much.
- ☆ While inserting the pipette, care should be taken that turbulence is not caused in thesuspension and pipetting is done uniformly at a moderate speed.
- ☆ To avoid interference among the particles, the concentration of the suspension should not exceed 2%. This means that not more than 20g of soil should be used.
- ☆ The diameter of the cylinder should be large enough to reduce the error due to the attraction of the walls of the vessel for soil particles. One-litre cylinder fulfills this condition.

- ☆ The dispersion of the soil particles through mechanical action (shaking and stirring) should be thorough and enough to get a uniform dispersion of individual particles.
- ☆ Pipetting should be done slowly so that the soil suspension is not disturbed.
- ☆ In both pipetting start the operation at least 20 seconds before the actual sampling time.
- ☆ Adjust your time schedule in such a way that you prepare the suspension on the first day and the first pipetting on the next day.

The reporting of results: Results should be expressed as follows from the textural class triangle. The area of an equilateral triangle is divided into 12 parts. Each part represents the textural class of soil. The textural designation is based on the international soil- particle- fraction classifications While using the diagram the points corresponding to the percentage of silt and clay present in the soil under consideration are located on the silt and clay sides of the triangle respectively. Lines are then projected inwards from these points in the first case parallel to the sand side. The compartment in which the lines meet gives the name of the textural class of the soil.

Result: As per the mechanical analysis of the given soil the composition of the soil separates is

Sand = --- percent; Fine sand= --- percent ; Silt= --- percent ; Clay ---percent and accordingly from the textural class triangle, the texture of the soil is ---------------- .

1.7 Preparation of Soil Saturation Paste and Extraction Of The Saturation Extract

Apparatus China dish of 250-ml capacity or greater and spatula

Procedure

Take 250 g soil in a china dish and add distilled water while stirring with a spatula. The soil-water mixture is consolidated from time to time during the stirring process by tapping the container on the workbench. At saturation, the soil paste glistens as it reflects light, flows slightly when the container is tipped and the paste slides freely and cleanly off the spatula for all soils except those with a high clay content. After mixing, the sample should be allowed to stand for an hour or more, and then the criteria for saturation should be rechecked. Free water should not collect on the soil surface nor should the paste stiffen markedly or lose its glistening appearance on standing. If the paste does stiffen or lose its glisten, remix with more water. Because soils puddle most readily when worked at moisture contents near field capacity, sufficient water should be added immediately to bring the sample nearly to saturation. If the paste is too wet, additional dry soil may be added.

Special precautions must be taken with peat and muck soils and with soils of very fine and very coarse texture.

Peat and muck soils: Dry peat and muck soils especially if coarse or woody in texture, require an overnight wetting period to obtain a definite endpoint as these soils usually stiffen and lose the glisten or standing. Adding water and remixing then gives a mixture that usually retains the characteristics of a saturated paste.

Fine textured soils: To minimize puddling and thus obtain a more definite endpoint with fine-textured soils, the water should be added to the soils with a minimum of stirring, especially in the earlier stages of wetting.

Coarse textured soils: The saturated paste for coarse-textured soils can be prepared in the same manner as for fine-textured soils; however, different moisture content is recommended for the salinity appraisal of such soils.

Extraction of Saturation Extract of Soil

Apparatus

Buchner funnels, filter paper, vacuum pump, extract containers such as test tubes

Procedure

Transfer the saturated soil paste, to a Buchner funnel with filter paper in place and apply a vacuum. Collect the extract in a bottle or test tube, Pyrex should not be used if boron is to be determined. If the initial filtrate is turbid, it can be refiltered through the soil or discarded. Vacuum extraction should be terminated when the air begins to pass through the filter. If carbonate and bicarbonate determination are to be made on the extract, a solution containing 1,000 ppm of sodium hexametaphosphate should be added at the rate of one drop per 25ml of extract prior to stoppering and storing. This prevents the precipitation of calcium carbonate on standing.

1.8. Soil Reaction

Principle

The pH value is a measure of the hydrogen or hydroxyl ion activity of the soil-water system and indicates whether the soil is acidic, neutral or alkaline in reaction. Since the growth of fish suffers much both under very low (strongly acidic) as well as high alkaline conditions, suitable reclamation measures become necessary. The presence of natural soluble salts as in saline soils is not normally reflected in the pH but when in large excess they tend to reduce H^+ activity.

Since the dissociation/ionic activity increases with temperature, the pH scale of 0 to 14 is valid only for a particular temperature (25°C). Thus all instruments are provided with a temperature setting/correction knob that must be made use of.

The pH of the soil is measured potentiometrically either in 1:5 saturated soil paste or I:2 soil-water suspension or 0.01 M $CaCI_2$ solution. The saturated paste method is used for identifying specific soil problems like acidity or alkalinity and salinity. For making fertilizer recommendations, the 1:2 soil-water suspension method is adopted. The pH in $CaCI_2$ solution is measured to mask the salt content variability in soils. The hydrogen ion concentration varies considerably with changes in the soil: water reaction and it is most important that the reaction adopted be standardized. Determination of soil reaction electrometrically falls into two distinctly separate parts viz. the preparation of soil: water suspension in which the pH is to be determined and the correct measurement of pH in a properly prepared soil: water suspension.

Apparatus

Shaking machine, pH meter with a combined electrode and electronic balance

Reagents

Standard buffer Solutions: These may be solutions of pH 4.0, 7.0, and 9.2 in pure water or of other pH values in the acidic and alkaline range. In the case of buffer tablets (available commercially), a single piece is to be dissolved in freshly prepared double distilled water and made up to 100 ml. It is advisable to prepare a fresh buffer solution after every few days as the solutions do not stay for long. A 0.05 M solution of AR grade potassium hydrogen phthalate ($KHC_8H_4O_4$ Mol. weight 204.14) gives a pH of 4.0 (actually 3.97) at 25 °C and this can also be used as a standard buffer. A stock solution of 0.25 M is prepared by dissolving 51.04 g of the pure dry salt in warm water and making up to 1 litre. Three to four drops of toluene are added to prevent the growth of mold. A dilution of 10 ml of this solution to 50 ml with water gives 0.05 M solution. Similarly, a solution containing 19.45 ml of 0.2 M NaH_2PO_4 and 0.55 ml of 0.1 M citric acid makes a buffer solution of pH 8.0.

Procedure

pH in 1:5 saturated soil paste: This is prepared by adding small amounts of distilled water to 100 g of soil in a 500 ml beaker while mixing with a spatula. At saturation, the soil paste glistens, flows slightly when the container is tilted, and slides freely and cleanly off the spatula. After mixing, it should be allowed to stand with a cover above the container for about an hour. Free water must not accumulate on the soil surface nor should be paste stiffen markedly or lose its glistening appearance on standing. The pH meter is to be set for the temperature and calibrated with two buffers, one in the acidic side and the other in the alkaline or neutral range. The electrodes (glass and calomel or combined) are carefully inserted in the paste and the pH measurement is made. The electrode is moved a little to ensure the removal of water film around it and pH reading is again taken when it is constant.

Soil to water ratio of 1:2; 20g of soil is taken in a 100 ml beaker to which 40 ml of distilled water is added. The suspension is stirred at regular intervals for 30 minutes and the pH is recorded. The suspension must be stirred well just before the electrodes are immersed and the reading taken.

pH in 0.01 M $CaCI_2$ solution: To 20 g of soil in a beaker, 40 ml of 0.01 M $CaCI_2$ solution (1.47 g AR grade $CaCI_2$. $2H_2O$/ liter) is added. The suspension is shaken thoroughly (30 minutes) and the pH is recorded. The reading is shifted downwards in comparison to the pH value obtained in water suspension. With acid soils, it normally ranges from 4 to 5, and in calcareous ones between 6.9 and 7.2. The pH of near-neutral soil in $CaCI_2$ solution is usually around 6.

N.B: Before each determination, the electrodes must be washed with a jet of distilled water and water is removed from the surface with a piece of filter, tissue, or blotting paper.

1.9 Electrical Conductivity

Principle

Since ions are the carriers of electricity, the electrical conductivity (EC) of the soil water system rises according to the content of soluble salts in the soil, giving rise to more of ion pairs on dissociation as it happens in the case of a dilute solution. Thus the measurement of EC can be directly related to the soluble salt concentration of the soil at any particular temperature. Conductivity of a solution is a measure of the total quantity of dissolved ions.

Instrument

The apparatus for measuring electrical conductivity consists of an AC salt bridge or electrical resistance bridge and a conductivity cell having electrodes coated with platinum black. The instrument is also available as an already calibrated assembly (solubridge) for giving directly the conductivity of solutions in millimhos per cm at 25 C

Reagent

0.0I M potassium chloride solution: 0.7456 g of dry (by drying a small quantity of AR grade potassium chloride at 60 C for 2h.) potassium chloride is dissolved in freshly prepared double distilled water and made to one liter. At 25°C it gives an electrical conductivity of 1411.8 x 10^{-6} (0.001418) mhos/cm. or 1.41 mmhos/cm. The conductivity bridge is to be calibrated and cell constant determined with the help of this solution. Even if the scale is marked directly in mmhos/cm (as in solubridge) it is necessary to check and calibrate the instrument with the KCI solution.

Procedure

20 g of soil is shaken intermittently with 40 ml of distilled water in a 150 ml Erlenmeyer (conical) flask for one hour and allowed to stand. Alternatively the clear

extract after pH determination can be used for electrical conductivity measurement. The conductivity of the supernatant liquid is determined with the help of the salt (conductivity) bridge. The measurement of EC (expressed in mmhos/cm) is adjusted for a known temperature (usually 25 °C) of the solution by setting the knob provided for this purpose.

1.10 Determination of the Cation Exchange Capacity (CEC) of Soil

Principle

It is the capacity of the soil to exchange cations for others on its exchange complex. It is usually expressed as meq per 100 gram of soil or Cmol (P+) kg^{-1} of soil. For CEC determination, the soil is leached with the suitable quantity of neutral N Ammonium acetate. Exchangeable cations are displaced by NH_4^+ ions. The excess ammonium acetate (NH_4OAC) is washed with alcohol. The amount of NH_4^+ ions adsorbed is determined by distillation with MgO.

Apparatus

1. Beaker (400 ml)
2. Burette
3. Pipette
4. Conical flak
5. Funnel
6. Kjeldahl flask
7. Distillation apparatus etc.

Reagents

1. Neutral 1 N Ammonium acetate solution (pH 7.0) : Dissolve 154 g ammonium acetate in distilled water, and dilute it to 1.8 litres. Mix thoroughly. Adjust pH to 7.0 with dilute ammonium hydroxide or acetic acid as required and make volume to 2 litres.
2. 40% alcohol: Take 40 ml of absolute alcohol (ethyl alcohol) and makeup volume 100 ml with distilled water.
3. MgO powder
4. N/10 Sulphuric acid solution: Take 2.8 ml of concentrated H_2SO_4 (36N) with an automatic pipette and dilute to one litre with distilled water.
5. N/10 Sodium hydroxide: Dissolve 4.0 gram of sodium hydroxide (solid) in distilled water and makeup volume to 1 litre.

6. Methyl red indicator.
7. Ammonium chloride (NH_4Cl) solution.

Procedure

1. Take 20 grams of soil in a 400 ml beaker and add 100 ml of 1 N ammonium acetate solution, stir well and let it stand overnight, then filter the contents of the beaker through Whatman No. 44 filter paper.
2. After the solution has leached down, add a drop of NH_4Cl solution and wash the soil with 25 ml of 40% alcohol, at each washing.
3. Continue this washing till the filtrate becomes chloride free (which can be tested with $AgNO_3$ solution).
4. After complete washing, transfer the soil along with the filter paper in Kjeldahl flask.
5. Add about 2-3 grams of MgO powder and 250 ml of distilled water and connect the flask to the distillation assembly.
6. Take 25 ml of N/10 sulphuric acid solution and 3-4 drops of methyl red indicator in a 250 ml conical flask, then put this conical flask below receiver of the distillation apparatus.
7. Collect distilled ammonia in this acid solution till the volume of distillate becomes 125 ml.
8. Titrate the excess acid with N/10 NaOH solution and calculate the C.E.C.

Observations

1. Weight of the soil taken = 20 g
2. Volume of N/10 H2SO4 solution taken (x) = ---------- ml.
3. Volume of N/10 NaOH solution used (y) = ----------- ml.
4. Volume of N/10 H2SO4 solution used (x-y) = ----------- ml.

Calculations

$$\text{Cation exchange capacity (CEC)} = \frac{\{(x\text{-}y) \text{ X } 0.018\}}{10} \times \frac{100}{20} \times \frac{1000}{18}$$

Result :Cation exchange capacity (CEC) of soil is ---------------- Cmol (P+) kg^{-1} of soil.

II. Soil was mixed with an excess of sodium acetate solution. Then it was washed with ethanol and then it was replaced with ammonium acetate. Final sodium concentration was estimated in flame photometer (Chapman, H.D., 1965).

1.11 Soil Organic Carbon

Principle

The soil is oxidized with a mixture of potassium dichromate and concentrated sulphuric acid and the organic carbon in the soil sample is oxidized utilizing the heat of dilution of sulphuric acid. The excess potassium dichromate is then titrated with ferrous sulphate or ferrous ammonium sulphate (redox titration). Organic Carbon in the sample is oxidised as follows:

$$2H_2Cr_2O_7 + 3C + 6\ H_2SO_4 = 2Cr_2\ (SO_4)_3 + 3CO_{2+}\ 8H_2O$$

Titration Method (Walkley and Black, 1934)

Reagents

Standard I N $K_2Cr_2O_7$ solution (Dissolve 49.4 g $K_2Cr_2O_7$ in water and dilute to 1 liter).

Standard 0.5 N Ferrous ammonium sulphate solution (Dissolve 196.1 gm of Ferrous ammonium sulphate in about 800 ml water containing 20 ml concentrated H_2SO_4 and dilute to 1 litre).

Diphenylamine indicator (Dissolve 0.5 reagent grade diphenylamine in 20 ml water and 100 ml concentrated H_2SO_4).

Orthophosphoric acid (85%) or Sodium fluoride

Conc. sulphuric acid

Procedure

For this determination, the soil sample should be grounded to pass through a 0.5 mm sieve. Place 1.0 g of the prepared soil sample in a 500 ml conical flask. Add 10 ml of 1 N $K_2Cr_2O_7$ solution and 20 ml of concentrated H_2SO_4. Mix thoroughly and allow the reaction to complete for 30 minutes. Dilute the reaction mixture with 200 ml water and 10 ml H_3PO_4. Add 10 ml of NaF solution and 1 ml of diphenylamine indicator. Titrate the solution with the 0.5 N ferrous ammonium sulphate to a brilliant green colour. A blank without soil is run simultaneously. If the titre value is <6, repeat the determination by taking 0.2 to 0.5 g of soil sample.

Calculation

One ml of N potassium dichromate is equivalent to 3 mg of carbon.

Organic carbon in soil is given by the following equation:

$$\text{Organic Carbon (\%)} = \frac{10\ (\text{B-V}) \times 0.003}{\text{B}} \times \frac{100}{\text{Weight of soil (g)}}$$

B = Volume of ferrous ammonium sulphate solution required for blank

V = Volume of ferrous ammonium sulphate solution required for the sample.

0.003 = Conversion factor

Note:

(i) Use of NaF along with H_3PO_4 gives a sharp end point.

(ii) High chloride content, as in the case of saline soils, interferes with the estimation. It can be prevented by adding Ag_2SO_4 @ 1.25 percent to the conc. Sulphuric acid.

Colorimetric Method (Datta et al., 1962)

Principle

The intensity of the green colour of chromic acid obtained due to its reduction is measured colorimetrically. The intensity is directly proportional to the amount of organic carbon present in the soil.

Instrument

Colorimeter or spectrophotometer and electronic balance

Reagents

1N potassium dichromate (Dissolve 49.4 g of K_2Cr_2O7 in distilled water and make it 1 liter)

Anhydrous sucrose (AR grade / Excelar)

Conc. Sulphuric acid

Procedure

Transfer I g of soil sample into a 100 ml conical flask. Add 10 ml 1N $K_2Cr_2O_7$ and 20 ml conc.

H_2SO_4 Swirl a little and keep on an asbestos sheet for 30 minutes. Decant about 15 to 20 ml of supernatant liquid into a centrifuge very carefully, retaining soil particles in the flask as far as possible. Centrifuge for about 5 minutes to ensure the complete settling of soil particles, if any.

Carefully decant the clear liquid into the colorimeter/spectrophotometer cuvette and read the colour intensity using a red filter / 660 nm wavelength.

Simultaneously run a blank without soil and a series of standards as follows. Accurately weigh 5, 10,15, 20, and 25 mg of anhydrous sucrose crystals into a 100 ml conical flask. Repeat as above for oxidation of sucrose and colour intensity development. Plot the colorimeter / spectrophotometer readings against the quantity of sucrose or carbon. Find out a factor to be multiplied by the sample reading to get the organic carbon content (%) of soil. The average value of this factor comes to be 0.0042 with little variation.

Calculation

Organic Carbon (%) = Colorimeter reading X 0.0042.

1.12 Total Nitrogen in Soil

Principle

Nitrogen is determined in soils by one of the several modifications of Kjeldahl's method in which the organic matter is oxidized by sulphuric acid and the nitrogen is converted to ammonia. The various modifications differ from the original in the addition of potassium or sodium sulphate to raise the temperature of the digest and in the use of other catalysts like selenium or copper sulphate in place of mercury. The sample is digested in a mixture of sulpuric acid and selenium.

Apparatus

Kjeldahl digestion unit, dispenser for 10 ml and burrette

Reagents

- ☆ Potassium sulphate or anhydrous sodium sulphate
- ☆ Selenium, granulated zinc, Sodium hydroxide
- ☆ Phenolphtahalein, Hydrochloric acid, Conc. Sulphuric acid
- ☆ Methyl red, Boric acid, and Bromo phenol blue.

Procedure

Take 10 g of soil in a flat-bottomed Kjeldahl digestion flask and add 10 ml of water. Shake and allow to stand for half an hour. Then add 30-35 ml of conc. sulphuric acid. Start the digestion, over a small flame and gradually increase the heat until white fumes of sulphuric acid are produced. Remove the flask and add 10 g of potassium sulphate (or anhydrous sodium sulphate) and 0.2 g of selenium. Replace the flask and continue the digestion for 1 to 2 hrs. after the digest becomes colourless. Allow the contents to cool, dilute with about 100 ml of water and transfer the fluid part to a 1000 ml conical flask, leaving as much as possible of the sand behind. Wash the sandy residue with four or five lots of 50-60 ml of water, decanting the washings into the conical flask after allowing the sandy residue to settle for a few seconds each time. Add a piece of granulated zinc and then add 100 – 110 ml of 40% sodium hydroxide solution.

Pour the sodium hydroxide solution down the side of the flask so that it forms a heavy layer at the bottom. Place the stopper in the flask and connect it to the distillation apparatus. Mix the contents well by shaking and commence the distillation, collecting the ammonia in 25 ml of 0.1N hydrochloric acid containing two or three drops of methyl red indicator solution. For soils high in nitrogen, absorb the ammonia in 35 ml of acid or take a smaller quantity of soil for the determination. Distil until about one-third of the liquid has passed over. When the distillation is complete, rinse the condenser tube that dips into the standard acid and titrate the excess acid with one-tenth of normal sodium hydroxide. Carry out a blank determination in exactly the same manner but using about 0.2 g of sucrose in place of the soil, so as to correct for any nitrogen contained in reagents.

Calculation

The percentage of nitrogen in the = (T-B) X N X 0.14

The soil on the basis of a 10 g sample

Where,

B = Blank titration, in ml. of standard alkali.

T = Actual titration, in ml. of standard alkali.

N = Normality of standard alkali.

Winkler's Modification

Procedure

Proceed as before but collect the distillate in 50 ml of 4 percent boric acid solution instead of standard hydrochloric acid. When the distillation is complete, add 3 drops of bromo phenol blue indicator solution and titrate the ammonium absorbed with 0.I N hydrochloric acid

Calculation

The percentage of nitrogen in the soil, on the basis of 10 g sample

= (T-B) X N X 0.14

where,

T = Volume of standard acid used in actual titration.

B = Volume of standard acid used in blank titration.

N = Normality of the standard alkali.

Nitrates and Ammonia

Nitrates and nitrites are readily soluble in water and the amount extracted is independent of the soil : water ratio. Ammonia, on the other hand, is present as an exchangeable ion. It is adsorbtively bound to the exchange complex and it must, therefore, be replaced by an excess of some other ion if it is desired to obtain it in solution quantitatively. Nitrites are seldom present in soil in amounts sufficient to warrant their determination, except in special cases. In most soils, they are so rapidly oxidized to nitrates that it is doubtful whether reliable values are obtained, even when the determination is carried out immediately after sample collection.

1.13 Nitrate-Nitrogen

Principle

Nitrates in the soil can be easily extracted with water along with a pinch of $CaSO_4$. $2H_2O$ or CaO. Clear soil extract may also be obtained by using $CuSO_4$, $Ca(OH)_2$ and $MgCO_3$. Interference of chloride ions can be prevented by the addition of silver sulphate to the $CuSO_4$ solution. NO_3^-N in the extract can be determined colorimetrically or by reducing with the help of Devard's alloy followed by distillation. The colorimetric method is based on the nitration of phenol disulphonic acid in fuming sulphuric acid as shown below:

$$C_6H_3(OH)(HSO_3)_2 + HNO_3 \rightarrow C_6H_2O_4(HSO_3)_2NO_2 + H_2O$$

The colourless product (in acid) behaves like nitrophenolic type indicator. It turns yellow when neutralized or made alkaline with NH_4OH due to the formation of nitrophenol disulphonic acid.

Instrument

Colorimeter or spectrophotometer and electronic balance

Reagents

Extraction solution: Mix 20 ml of 0.5 M $CuSO_4$ solution and 100 ml of 6% silver sulphate solution, dilute to 1 litre with distilled water and mix well.

Phenol disulphonic acid: Weigh 25 g of AR grade phenol (white crystals) or 25 ml of pure liquid phenol in a round botton or Kjeldahl flask. Add 250 ml of NO_3 free conc. H_2SO_4 while shaking the flask. Add 75 ml of fuming sulphuric acid and mix well. Place the flask in boiling water dipping it up to the height of the contents in the flask. Heat for 2 hrs, cool and store the resultant phenol disulphonic acid in a glass stoppered bottle. In case fuming sulphuric acid is not available, the same amount of conc. Sulphuric acid may be added and boiled for 6 hrs.

Dilute ammonia solution: Conc. Ammonia solution (sp. Gravity 0.91) diluted with equal amount of distilled water

Standard nitrate solution: Weigh 0.7215 g of AR grade, oven-dried and cooled KNO_3 and add little water to dissolve and make the volume to 1 litre. This stock solution contains 100 mg L^{-1} nitrate-nitrogen. From this, prepare a working solution containing 10 mg L^{-1} NO_3^-N.

$Ca(OH)_2$ (AR Grade)

$MgCO_3$ (AR Grade)

Procedure

Place 5 g of soil sample in a 100 ml conical flask. Add 25 ml of the extracting solution and shake for 10 min. Add a pinch (about 0.2 g) of $Ca(OH)_2$ and shake for 5 minutes. Add about 0.5 g of $MgCO_3$ and again shake for 10 minutes. Allow to settle for about 5 minutes and filter through Whatman No. 42 filter paper, discarding the first few ml of filtrate. Transfer 10 or 15 ml of the clear and colourless filtrate to a 50 ml porcelain/silica dish or 100 ml beaker. Work in NO_3—free atmosphere. Evaporate the contents to dryness in a water bath. Do not heat beyond dryness. Cool at room temperature. Rapidly add 3 ml of phenol disulphonic acid, covering the entire residue by rotating the dish/beaker. After 10 minutes add 15 ml of distilled water and stir with a glass rod. Cool and transfer the contents quantitatively to a 100 ml volumetric flask. Add diluted (1:1) ammonia solution slowly, while mixing until the solution becomes alkaline. This is indicated by the development of yellow colour due to the presence of nitrate. Add another 2 ml of diluted ammonia solution and make the volume 100 ml with distilled water. Measure the intensity of the yellow colour using a blue filter (420 nm).

Preparation of standard curve

Transfer 2, 5, 10, 15, 20, and 25 ml of 10 mg $L^{-1}NO_{3^-}$ N solution into dishes/beakers and evaporate to dryness. Run a blank without an aliquot, develop yellow colour as described above, and read the colour intensity in an identical manner. Draw a standard curve by plotting the concentration of NO_3^-N against readings.

Calculation

NO_3^--N (mg / 100g) =

$$\text{Mg of } NO_3^-\text{-N in aliquot} \times \frac{25}{vol.\ of\ aliquot} \times \frac{100}{wt.\ of\ soil(g)} \times \frac{1}{100}$$

1.14 Ammoniacal Nitrogen

Principle

Like metallic cations, ammonium ions are also held in exchangeable form in soils. In the interlayer positions of the 2:1 layer silicates, NH_4^+ -N is usually extracted with NaCI or KCI solution. This also extracts the NO_{3^-}N which is water soluble. However, nitrites being negligible may be ignored. NH_4^+-N in the NaCI or KCI extract may be determined by distillation or by using a colorimeter or spectrophotometer. The colorimetric method (Nessler's method) is suitable for samples having less than 10 mg N kg^{-1} soil.

Calorimetric method

Instruments

Colorimeter or spectrophotometer and mechanical shaker

Reagents

10% Sodium chloride (acidified) : Dissolve 100 g of NaCI per liter and adjust the pH at 2.5 with HCI.

Nessler's reagent: Dissolve 45.5 g of mercuric iodide and 35.0 g of potassium iodide in small amount of water in a beaker. Transfer it into a one liter volumetric flask. Add 112 g of KOH and bring the volume to about 800 ml. Mix well, cool, and make up the volume to one liter. Allow to stand for a few days and decant the supernatant liquid into an amber coloured bottle.

10% Sodium tartrate: Dissolve 100 g $Na_2C_4H_4O_6$ $2H_2O$ in distilled water and make up the volume to one liter.

Standard NH_4CI solution: Dissolve 1.337 g of pure NH_4CI in water and make up the volume to one liter to get 0.025 me NH_4^+ ml^{-1}. Dilute suitable volume of it 25 times to get a working solution containing 0.00I me of NH_4^+ ml^{-1}.

Procedure

Place 100 g of fresh soil in a 500 ml conical flask. Determine moisture content separately by drying at 105 °C in an oven. Add 200 ml of acidified NaCI solution and shake for 30 minutes. Fit an 11 cm size Whatman No. 42 filter paper circle in a Buchner funnel, moisten and fix it firmly by a mild suction through electric suction pump or a simple glass-made suction device fitted to a water tap. Pour the contents into the Buchner funnel to filter. Add 25 ml of the NaCI solution in the conical flask to wash and rinse out and transfer to the funnel. Add 8 to 9 portions of 25 ml of NaCI solution each and collect the entire leachate. Transfer the leachate into a 500 ml volumetric flask and make up the volume. Take 50 ml aliquot into a 100 ml volumetric flask and add 2 ml each of the tartrate and acidified NaCI solutions. Add a pinch of gum acacia as a protective colloid and bring the volume to about 93 ml. Add 5 ml of Nessler's reagent with rapid mixing. Make the volume to 100 ml. Measure the intensity of the orange colour developed, after 25 minutes at 410 nm (blue filter).

Preparation of standard curve for NH_4^+-N

Place 3 to 30 ml portion of the working standard solution containing 0.00I me of $NH_4^+ml^{-1}$ in a series of 100 ml volumetric flasks. Proceed for colour development as in the case of sample aliquots. Measure the colour intensity and draw a standard curve by plotting colorimeter readings against the NH_4^+ concentrations.

Calculation

NH_4^+ (me/100 g soil) on dry wt. basis.

$$= \frac{R \text{ x volume of extractant used}}{\text{Volume of aliquot taken}} \times \frac{100}{(100-M)}$$

Where,

R stands for the NH_4^+ in me in the final volume i.e. 100 ml as read from standard curve, and M stands for moisture percentage in soil sample. The results may then be expressed on fresh weight basis also.

1.15 Available Nitrogen

Principle

The easily oxidisable nitrogen is estimated by Subbiah and Asija (1956) method. The procedure involves distilling the soil with alkaline $KMnO_4$ which oxidizes and hydrolyses the organic matter present in the soil. The liberated ammonia is condensed and absorbed in boric acid, which is titrated against standard acid. This serves as an index of the available N status of soil and is therefore proposed as a soil test for N. The method has been widely adopted to get a reliable index of nitrogen availability in soil due to its rapidity and reproducibility. The process of oxidative hydrolysis is, however, a progressive one, and hence a uniform time and heating temperature should be allowed for the best results. The use of glass beads checks to bump while liquid paraffin checks frothing during heating.

Apparatus

Nitrogen distillation unit preferably with six regulated heating elements.

Reagents

0.32%$KMnO_4$: Dissolve 3.2 g of $KMnO_4$ in water and make to one liter.

2.5%NaOH: Dissolve 25 g of sodium hydroxide pellets in distilled water and make the volume to one liter.

2% Boric acid: Dissolve 20 g of boric acid powder in warm water by stirring and dilute to one liter.

Mixed indicator : Dissolve 0.066 g of methyl red and 0.099 g of bromocresol green in 100 ml of ethyl alcohol. Add 20 ml of this mixed indicator to each liter of 2% boric acid solution. Adjust the pH to 4.5 with dil. HCI or dil. NaOH.

0.1N Potassium hydrogen phthalate: Dissolve 20.422 g of the salt in water and dilute to one liter. This is a primary standard and does not require standardization.

0.1 N NaOH: Dissolve 4 g of NaOH in water and dilute to one liter. Standardize it against 0.1N Potassium hydrogen phthalate solution.

0.02N H_2SO_4: Prepare approximately 0.1N H_2SO_4 by adding 2.8 ml of conc. H_2SO_4 to about 990 ml of distilled water. From this, prepare 0.02N H_2SO_4 by diluting a suitable volume five times with distilled water. Standardize it against 0.IN NaOH solution.

Procedure

Place 20 g of soil sample in an 800 ml Kjeldahl flask. Moisten the soil with about 10 ml of distilled water, and wash down the soil adhering to the neck of the flask, if any. Then add 2 ml of liquid paraffin and a few glass beads to prevent frothing and bumping respectively. Add 100 ml of 0.32% $KMnO_4$ solution. Measure 20 ml of 2% boric acid containing mixed indicator in a 250 ml conical flask and place it under the receiver tube. Dip the receiver tube end in the boric acid. Run tap water in the condenser. Add 100 ml of 2.5% NaOH solution and immediately attach to the rubber stopper fitted in the alkali trap. Switch the heaters on and continue distillation until about 100 ml of distillate is collected. First, remove the conical flask containing distillate and then switch off the heater to avoid back suction. Titrate the distillate against 0.02N H_2SO_4 taken in a burette until pink colour just appears. Run a blank without soil with each set of five samples. Carefully remove the Kjeldahl flask after cooling and drain the contents in the sink.

Precautions

Check all the joints of the Kjeldahl apparatus to prevent any leakage and loss of ammonia. Hot Kjeldahl flasks should neither be washed with cold water nor allowed to cool for long to avoid deposits at the bottom. In case frothing takes place and reaches the trap, the samples should be discarded and fresh distillation should be done. Opening ammonia bottles in the laboratory should be strictly prohibited while distillation is on. The titration should be carried out in ammonia-free atmosphere. In case the titration is not to be carried out immediately, the distillate should be stored in ammonia-free cupboards in tightly stoppered flasks.

Calculation

$$\text{Available N (mg/100g)} = \frac{\left[(S-B)\ x\ (0.28)\right]}{Weight\ of\ \text{the}\ soil} \times 100$$

Where,

S and B stand for the titer value of the sample and blank respectively.

1.16 Determination of Total Nitrogen in Soil and Plant

Total nitrogen is estimated by the micro-Kjeldahl method as per the procedure suggested by AOAC (1995). Plant analysis has been considered a superior diagnostic technique for mineral content. The whole plant is dried in the open air for a few days after that it was further dried in a hot air oven at about 60 ± 2° C for eight to ten hours per day to achieve complete drying. After drying, whole plant is powdered with the help of a grinder, passed through a 2 mm stainless steel sieve, and used for the chemical assay. The soil sample from a definite depth was randomly collected from the field with the help of a screw auger. All the possible technical precautions as prescribed for standard soil sampling were also taken. Samples were brought to the laboratory, air-dried in the shade, and grounded by a wooden roll, thereafter sieved through a 2 mm stainless steel sieve and stored in polythene bags, and used for the chemical assay.

Principle:

Nitrogen in samples like plants and soil exists in a very complicated bonding structure. During digestion, a known weight of the plant/soil samples in the presence of sulphuric acid with catalyst mixture under high temperature is digested where complicated structures are broken to simple structures, thereby releasing nitrogen in the form of ammonium radical (NH_4^+). During distillation in presence of sodium hydroxide, the released ammonia is condensed and absorbed in known volume of a boric acid with a mixed indicator to form ammonium borate, the excess of which is titrated with standard sulphuric acid.

The micro-Kjeldahl method consists of three steps;

1. Digestion
2. Distillation and
3. Titration.

1. KEL PLUS Automatic Nitrogen Estimation System

The said instrument is used for determination of nitrogen. It consists of the following;

- ☆ **Macro Block Digestion System (Model KES 12L):** This digestion system is suitable for soil, plant, water, pesticides, fertilizers, food and feed samples. It is microprocessor based automatic twelve place macro block digestion system with temperature controller fitted with sensor break protection (Microprocessor based) feature and temperature range from 50-450 °C.
- ☆ **Acid Neutralizer Scrubber (Model KEL VAC):** It is used to neutralize the acid fumes which are absorbed in 15% sodium hydroxide and dissolved

in water stored in the system tank. After every 2 cycles of digestion, the 15% sodium hydroxide solution is replaced and after 3 cycles of digestion, acid fumes dissolved in water tank are drained off and refilled with fresh water in the system tank.

- ☆ **Automatic Distillation System (Model Classic DX):** It is fully automatic distillation system with programmable auto run digital features, with automatic dilution and addition of boric acid and NaOH. Both modes (auto and manual) are available for distillation reagents addition.
- ☆ **Refrigerated Water Cooling System for Condenser (Model Kel Freeze) :** It is refrigerated water cooling system for distillation and condensing system with inbuilt compressor and recirculation pump.
  - ☆ Electronic balance
  - ☆ Burette
  - ☆ Pipette
  - ☆ Conical flask
  - ☆ Measuring cylinder
  - ☆ Distilled water

Reagents:

1. Concentrated sulphuric acid (H_2SO_4).
2. Catalyst mixture: Mix with 250 g potassium sulphate (K_2SO_4), 50 g cupric sulphate ($CuSO_4$. 5 H_2O) and 5 g metallic selenium powder in the ratio of 50:10:1.
3. 40 % sodium hydroxide (NaOH).
4. 4 % boric acid containing 20 - 25 ml mixed indicator /liter.
5. Mixed indicator: 0.066 g methyl red + 0.099 g bromocresol green dissolved in 100 ml of 95 % alcohol.
6. 0.02N sulphuric acid (H_2SO_4).

Procedure /Digestion

- ☆ Weigh 0.5 g of prepared plant sample or 1 g of soil sample and transfer it to the digestion tube
- ☆ Add 10 ml of concentrated sulphuric acid and 5 g of catalyst mixture to the sample
- ☆ Load the digestion tubes in to the digester and heat the digestion block. Switch on the digestion unit and set the initial temperature 100 ^{0}C till frothing is over.

- ☆ Then block temperature is raised to 400 °C. The effective digestion starts only at 360 °C and beyond 410 °C.
- ☆ The sample turns light green colour or colourless at the end of the digestion process

Distillation

- ☆ After cooling the digestion tube, load the tube in distillation unit and on other side of hose keep 20 ml of 4 % boric acid with mixed indicator in 250 ml conical flask.
- ☆ 40 ml NaOH (40 %) is automatically added by distillation unit programme.
- ☆ The digested sample is heated by passing steam at a steady rate and the liberated ammonia absorbed in 20 ml of 4 % boric acid containing mixed indicator solution kept in a 250 ml conical flask.
- ☆ With the absorption of ammonia, the pinkish colour turns to green. Nearly 150 ml of distillate is collected in about 8 minutes. Simultaneously, blank sample (without plant/soil) is to be run.

Titration :

- ☆ The green colour distillate is titrating with 0.02N sulphuric acid and the colour changes to original shade (pinkish color).
- ☆ Note the blank & sample titer reading (ml) and calculate the total nitrogen content present in plant/soil samples.

Calculations :

Nitrogen content in plant (%) = { R (sample titer - blank titer) x Normality of acid x Atomic weight of nitrogen x 100} / (Sample weight (g) x 1000}

= (R x 0.1 x 14 x 100)/ (0.5*1000)

Factor = R x 0.28

Nitrogen content in soil (%) = = (R x 0.1 x 14 x 100)/ (1.0*1000)

Factor = R x 0.14

Crude Protein Content

The total nitrogen is estimated by the micro-Kjeldahl method as per the procedure suggested by AOAC (1995) and the crude protein is calculated by the following formula:

Crude protein content (%) = micro-Kjeldahl nitrogen content (%) x 6.25 (based on the assumptions that nitrogen constitutes 16 % of protein).

1.17 Digestion of the Sample (Plant, Fish, Feed) for Total Nutrient (N, P, K,S, Fe Estimation)

Take one gram of material in the digestion flask. Add 10-15ml of Diacid (3:1: Nitric acid: perchloric acid) mixture and swirl the contents in a digestion tube / flask or conical flask, or Triacid (100:10:40ml i.e 10:1:4 of HNO_3: H_2SO_4 : $HClO_4$). Place the contents on a hot plate till the digestion is over (white colour). Transfer the contents by filtering in 100 ml volumetric flask by rinsing the tube several times with hot water and passing through filter paper. Make the volume to 100ml with hot distilled water and store the solution in airtight container. (Note: For S estimation Diacid mixture only)

1.18 Carbon and Nitrogen Ratio (C:N ratio)

C:N ratio of the solid and liquid sample (fish, soil, feed, leaf, organic matter, water etc.) can be determined by using CHNS analyser. The carbon to nitrogen (C/N) ratio is significant in composting because microorganisms need a good balance of carbon and nitrogen (ranging from 25 to 35) in order to remain active. High C/N ratios can lead to prolonged composting duration and low C/N ratios enhance nitrogen loss. The carbon-to-nitrogen ratio is written as C: N and is usually a single number (Flavel and Murphy, 2006). Consequently, a ratio of 20 means that there is 20 g of carbon for each 1 g of nitrogen in that organic matter. When an organic substrate has a C: N ratio between 1 and 15, rapid mineralization and release of N occurs, which is available for plant uptake. The lower the C:N ratio, the more rapidly nitrogen will be released into the soil for immediate crop use (Watson et al., 2002). A C:N ratio > 35 results in microbial immobilization. A ratio of 20–30 results in an equilibrium state between mineralization and immobilization. Soil microorganisms have a C: N ratio of around 8. They must acquire enough carbon and some nitrogen from the soil to maintain that ratio in their cells and have been found to do best on a "diet" with a C: N ratio of 24 (Howell, 2005).

1.19 Available Phosphorus

Principle

Phosphorus exists in the form of various types of orthophosphates in soil. A very small fraction of these is in the available form at a given time. The available phosphorus content of soil consists mainly of Ca-, Al- and Fe-P. In the neutral and alkaline soils particularly, Ca-P is the dominant fraction. Organic-P fraction is also in considerable amount, but is usually not included in the determination of available phosphorus. A large number of extracting reagents from Dyer's 1% citric acid to some of the multinutrient extractants, buffer solutions, acids, and chelating agents have been suggested for available phosphorus determination from time to time. However, no single extractant appears to be suitable for all types of soils.

Two types of extraction methods are more popularly adopted. Under acidic soil conditions, Bray's P-1 (or Bray No.1) method which involves soil extraction with a solution containing 0.03 N NH_4F and 0.025 N HCI is widely followed. The fluoride complexes Al and Fe in the soil thus release some bound P besides the easily acid-soluble P (largely Ca-P). This extractant is suitable for soils containing less than 2% calcite or dolomite because, in calcareous soils, carbonate quickly neutralizes the acid, resulting in less extraction of P.

The other and most widely used extractant is the 0.5M $NaHCO_3$ solution at pH 8.5. The reagent is most suitable for neutral to alkaline soils and is designed to control the ionic activity of calcium through solubility product of $CaCO_3$, thus extracting the most reactive forms of P from AI-, Fe- and Ca-phosphates. Phosphorus in the extract can be determined using a suitable method of colour development and measuring the colour intensity at an appropriate wavelength. The dilute acid-fluoride extractant easily removes easily acid soluble-P from phosphates bound to Al, Fe and Ca- phosphate in the extract which is determined colorimetrically as phosphor-molybdenum blue with stannous chloride.

Bray's P-1 Method (Bray and Kurtz, 1945)

Principle

The dilute acid-fluoride extractant easily removes easily acid soluble P from phosphates bound to AI, Fe and Ca. Phosphate in the extract is determined colorimetrically as phospho-molybdenum blue with stannous chloride.

Instruments

Mechanical shaker and colorimeter or spectrophotometer

Reagents

Bray's P-1 extractant: Dissolve 1.110 g of AR grade ammonium fluoride in one liter of 0.025 N HCI.

1.5% Dickman and Bray's reagent: Dissolve 15 g of AR grade ammonium molybdate in 300 ml of warm water, cool, and add 350 ml of 10N HCI. Make the volume one liter.

40% $SnCI_2$ stock solution: Weigh 10 g of stannous chloride in a 100 ml glass beaker. Add 25 ml of conc. HCl and dissolve by heating. Cool and store in an amber-coloured bottle, after adding a small piece of Zn metal (AR grade) to prevent oxidation. From this prepare a dilute $SnCI_2$ solution (0.5 ml diluted to 66 ml).

100 mg P L^{-1} stock solution: Weigh 0.439 g of AR grade KH_2PO_4 dried in an oven at 60 C for 1 h in a one-liter beaker, add about 500 ml of distilled water, and dissolve. Add 25 ml of approx. 0.7 NH_2SO_4 and make the volume to one liter.

2 mg P L^{-1} solution: Dilute a suitable volume of 100 mg P L^{-1} solution by 50 times to get 2 mg P L^{-1} solution.

Procedure

Place 5 g of soil sample in a 150 ml conical flask. Add 50 ml of Bray's P-1 extractant and shake for 5 minutes. Filter through Whatman No.1 filter paper quickly so as to collect the filtrate within 10 minutes. Transfer 5 ml aliquot into a 25 ml volumetric flask. Add 5 ml of ammonium molybdate solution, shake a little, and dilute to about 22 ml. Add 5 ml of diluted $SnCI_2$ (0.5 ml diluted to 66 ml), mix by shaking a little and make up the volume. Run a blank without soil under identical conditions. Measure the intensity of the blue colour developed, using 660 nm wavelength.

Calculation

$$\text{Available P (mg/100g soil)} = \frac{R \times V \times 100}{A \times S \times 1000} = \frac{R \times 50 \times 100}{5 \times 2.5 \times 1000} = (R/2.5)$$

Where,

R = µg P in the aliquot (obtained from standard curve).

Olsen's method (Olsen et al., 1954)

Principle

Sodium bicarbonate solution extracts some exchangeable or surface- adsorbed AI-P, Fe-P, Ca-P and other phosphates.

Instruments

Colorimeter or spectrophotometer, electronic balance and mechanical shaker

Reagents

0.5 M $NaHCO_3$: Dissolve 42 g of P-free sodium bicarbonate in about 500 ml of hot water and dilute to one liter. Adjust the pH to 8.5 using dilute NaOH solution or dilute HCI. Store in a glass or polyethylene bottle

Activated charcoal: Wash pure activated charcoal or commercially available Darco G-60 with acid to make P-free, even if having traces of P.

1.5% ammonium molybdate solution: Similar to the described in case of Bray's P-1 except that use 400 ml of 10N HCI instead of 350 ml per liter.

40% $SnCI_2$ solution : Same as in case of Bray's P-1.

100 mg L^{-1} P working solution: Same as in case of Bray's P-1.

Procedure

Transfer 2.5 g of soil sample to 100 ml conical flask. Add a pinch of Darco G-60 and 50 ml of Olsen's reagent (0.5 M $NaHCO_3$, pH 8.5). Shake for 30 minutes on a mechanical shaker. Filter through Whatman No. 1 filter paper. Transfer 5 ml of clear and colourless filtrate into a 25 ml volumetric flask. Gradually add 5 ml of ammonium molybdate solution containing 400 ml of 10N HCI per liter. Shake slowly and carefully to drive out the CO_2 evolved. When frothing completely ceases, add distilled water, washing down the sides, to bring the volume to about 22 ml. Add 1 ml of freshly diluted $SnCl_2$ solution, shake a little, and make the volume to 25 ml. Read the blue colour intensity at 660 nm (red filter). Run a blank without soil under an identical manner.

Preparation of Standard Curve for P

Take six 25 ml volumetric flasks. Pipette out 0, 0.5, 1.0, 1.5, 2.0, and 2.5 ml of 2 mg L^{-1} P solution. Add 5 ml of the extractant (Bray's or Olsen's). Add ammonium molybdate as for Bray's or Olsen's method and proceed to develop blue colour as described earlier. Measure blue colour intensity and draw a standard curve by plotting concentration of P in µg against colorimeter readings. If a straight line is obtained, find out a factor for each reading.

Calculation

$$\text{Available P (mg/100g soil)} = \frac{R \text{ x } volume\ of\ extract \times 100}{Wt.\ of\ soil\ taken \times 1000}$$

Where,

R	=	quantity of P in µg read on X-axis against a sample reading,
V	=	Volume of extracting reagent used (ml),
A	=	volume of aliquot used for colour development (ml) and
S	=	weight of soil sample taken (g).

Ascorbic acid Method (Watanabe and Olsen, 1965)

Principle

The major lacuna with Bray and Kurtz's method of blue colour development is that the colour starts fading soon and hence the intensity has to be measured immediately. A method based on the use of ascorbic acid given by Watanabe and Olsen (1965) and given below provides a more stable blue colour and is therefore, preferred over the former.

Reagents

Molybdate-tartarate solution: Dissolve 12 g of ammonium molybdate in about 250 ml of distilled water to get solution 'A'. Prepare solution 'B' by dissolving 0.291 g antimony potassium tartarate in 100 ml of distilled water. Prepare one liter of 5N H_2SO_4 and add solution 'A' and 'B' to it. Mix thoroughly and make the volume to 2 L with distilled water.

Ascorbic Acid Solution: Dissolve 1.056 g of ascorbic acid in 200ml of the molybdate-tartarate solution and mix well. Prepare it fresh as and when required.

p-nitrophenol Indicator: Dissolve 0.5 g of p-nitrophenol in 100 ml of distilled water.

5N H_2SO_4 : Carefully dilute 140 ml of conc. H_2SO_4 to 1 L with distilled water to get approx. 5N H_2SO_4.

Procedure

Pipette 5 ml of the Olsen's reagent into a 25 ml volumetric flask and add 2-3 drops of p-nitrophenol indicator. It develops yellow colur. Add known quantity of 5N H_2SO_4 drop by drop to acidify the Olsen's reagent to pH 5 at which the yellow colour will disappear. Note down the volume of 5N H_2SO_4 used. Place a 5 ml aliquot of the Olsen's extract of soil to a 25 ml volumetric flask and add the required quantity of 5N H_2SO_4 to bring it to pH 5.0. Dilute to 20 ml with distilled water. Add 4 ml of the ascorbic acid solution, make the volume to 25 ml and shake well. Wait for 10 minutes and then measure the colour intensity at 730-840 nm. Run a blank with the extracting solution (without soil). Prepare standard curve as described for Bray and Kurtz method and calculate amount of P present in mg/100 g soil.

Precautions

All glassware to be used for P determination should be cleaned with chromic acid followed by thorough washing to minimize contamination. In case of delay in measuring colour intensity, cover the samples, store properly and add $SnCI_2$ just before measurements. Colour intensity measurement should be carried out exactly 10 minutes after developing blue colour. In $SnCI_2$ method colour starts fading soon after development.

1.20 Exchangeable Potassium

Principle

The term available potassium (K) conventionally refers to exchangeable + water-soluble K. The exchangeable K constitutes the major portion of available K except in saline or saline-sodic soils. Available K or exchangeable K is usually determined in neutral normal ammonium acetate (N NH_4OAc) extract of soil. The extraction is carried out by shaking soil with extractant followed by filtration or centrifugation. The K is estimated by a flame photometer or AAS. In soils with appreciable amount of soluble K, this cation is estimated in a saturation extract (Jackson, 1958) and deducted from N NH_4OAc extractable K to obtain exchangeable K.

Instrument

Flame photometer, mechanical stirrer, pH meter and electronic balance

Reagents

Ammonium acetate (1.0N, pH 7.0): To 700 ml of distilled water, add 57 ml 99.5% glacial acetic acid (CH_3COOH) and then 69 ml of conc. ammonium hydroxide (NH_4OH). Dilute to a volume of 900 ml and adjust pH to 7.0 by the addition of more of 3N NH_4OH or 3N CH_3COOH and make up to 1 liter. Store in a pyrex or polypropylene bottle. Alternatively, dissolve 154 g ammonium acetate (CH_3COONH_4) in water and dilute to 1.8 liter. Mix thoroughly. Adjust pH to 7.0 with dilute NH_4OH or HOAc as required and make to 2 Liters.

Potassium chloride standard solution: Make a stock solution of 1000 ppm K by dissolving 1.908 g AR grade potassium chloride (dried at 60 °C for 1 h) in distilled water and diluting up to 1 liter. Prepare 100 ppm standard by diluting 100 ml of 1000 ppm stock solution to one liter with extracting solution.

Standard curve: Pipette 0, 5, 10, 15 and 20 ml of 10 ppm solution into 100 ml volumetric flasks and bring the volume to mark with extracting solution. The solutions contain 0, 5, 10, 15, and 20 ppm K respectively.

Procedure

Shaking and filtration can be done by the method of Schollenberger and Simon (1945) or shaking and centrifugation can be done as described by Knudsen et al. (1982).

(a) Shaking and filtration (Schollenberger and Simon, 1945)

Transfer 5 g soil in a 150 ml Erlenmeyer flask and pour in 25 ml of neutral N ammonium acetate (pH 7.0). Shake on a reciprocating shaker (180 + oscillations/min) for 5 minutes and immediately filter through Whatman No. 1 filter paper. First few ml of the filtrate may be discarded.

(b) Shaking and centrifugation (Knudsen et at., 1982)

Place 10 g of <2 mm air-dried soil (or use 5 g if the soil contains greater than 500 ppm K in a 50 ml centrifuge tube). Add 25ml of ammonium acetate (NH_4OAc), cover and shake the tube for 10 minutes. Centrifuge the tube at 2000 rpm for 10 min until the supernatant liquid is clear. Decant the supernatant liquid into a 100 ml volumetric flask. Make three additional extractions in the same manner. Dilute the combined extracts to 100 ml with NH_4OAc. Mix the solution and determine K in the extract by the flame photometer using K filter after necessary setting and calibration of the instrument (Use 0 and 20 ppm working K concentrations). Read similarly the different concentration of K in flame photometer and obtain the standard curve by plotting the reading against the different concentration of K.

Calculation

$$\text{Available K (mg/100 g soil)} = \frac{R \text{ x volume of extract } \times 100}{Wt. \text{ of soil taken } \times 1000}$$

Where,

R = ppm of K in the extract (obtained from standard curve)

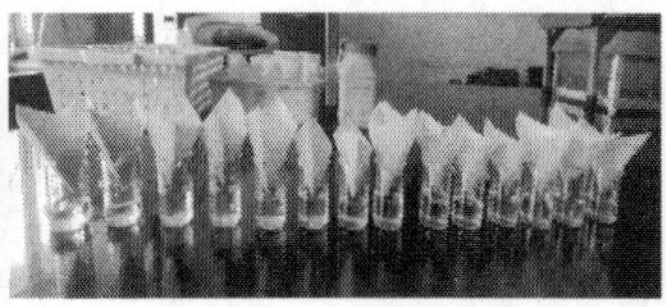
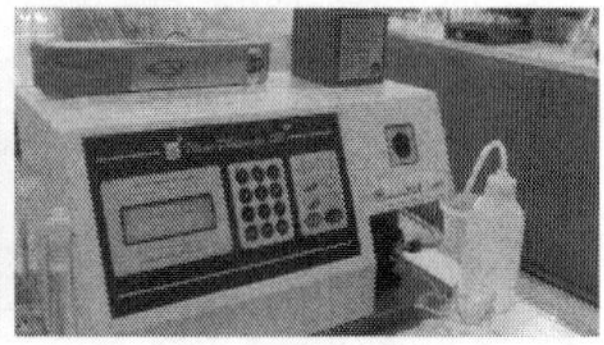

1.21 Available Sulphur

Available sulphur occurs in soil as soluble form as well as adsorbed sulphate ions. For estimating available sulphur in soil, 0.15% $CaCl_2$ extractable S method (Williams and Steinbergs, 1959) and monocalcium phosphate extractable-S (Chesnin and Yien, 1950) method are generally employed.

$CaCI_2$ Extractable S

In this method, S is extracted with 0.15% $CaCl_2$ and the soluble sulphate estimated turbidimetrically on a colorimeter using blue filter or spectrophotometer at 340 nm.

Reagents

Calcium chloride dihydrate ($CaCl_2.2H_2O$) 0.15%: Dissolve 1.5 g of $CaCl_2$. $2H_2O$ in about 700 ml of water and make to 1 liter with distilled water.

1.0% sodium chloride (NaCl): Dissolve 10 g of NaCl in 700 ml of distilled water and make to 1 liter with distilled water.

Barium chloride dihydrate: $BaCI_2$. $2H_2O$ crystals, 20 to 30 mesh.

Conditioning reagent: Dissolve 75g NaCI in 275 ml of distilled water in a 500 ml volumetric flask. Stirring with magnetic stirring bar, add 30 ml of conc. HCI, 100 ml of absolute ethanol and 50 ml of glycerol. Rinse glycerol into flask. Continue stirring until NaCI dissolves. Remove stirring bar and make the volume with distilled water.

Standard sulphate solution: Dissolve 0.5434 g K_2SO_4 in 1 liter of distilled water. This contains 100 mg S/ml. Dilute 10 times to obtain 10 ppm solution. Transfer 0, 5, 10, 15 and 20 ml (containing 0, 50, 100, 150 and 200 µg SO_4-S) to Erlenmeyer flasks of 150 ml, bringing the volume with distilled water to 30 ml.

Procedure Transfer 5 g soil into an 150 ml Erlenmeyer flask. Add 25 ml 0.15% $CaCI_2$ to it. Shake for 30 minutes on a reciprocating shaker (180 + oscillations/min). Filter the suspension through Whatman No. 42 filter paper. Pipette 10 or 20 ml extract in a 150 ml Erlenmeyer flask and add 20 or 10 ml of H_2O thus bringing the volume to 30 ml. Add 2.5 ml of stabilizing solution and 0.2 to 0.3 g of $BaCI_2$ crystals by spatula to the standards and extract. Shake the flasks for 1 minute each at constant rate. After 1 to 3 minutes, measure the turbidity in a colorimeter using a blue filter. Turbidity can be more accurately measured at 340 nm on a spectrophotometer. The turbidity remains constant for 3-10 min.

Calculation

$$SO_4\text{-S in soil mg/100 g soil} = \frac{Factor \times SampleR \times 1000 \times 100 \times 100}{1000 \times 10 \times 1}$$

Monocalcium Phosphate Extractable Sulphur

Principle

Phosphate ions as monocalcium phosphate are preferred for replacement of the adsorbed sulphate ions. Ca salts have an advantage over K or Na salts as Ca prevents deflocculation in heavy textured soils and helps in easy filtration. Sulphate in the extract is estimated turbidimetrically.

Instrument

Colorimeter or spectrophotometer or autoanalyzer, electronic balance and mechanical shaker

Reagents

Gum acacia-acetic acid solution: Dissolve 5 g of chemically pure gum acacia powder in 500 ml of hot water and filter in hot condition through Whatman No. 42 filter paper. Cool and dilute to one liter with dilute acetic acid.

Mono-calcium phosphate extracting solution (500 mg P L^{-1}): Dissolve 2.035 g of Ca $(H_2PO_4)_2.H_2O$ per liter.

Barium chloride: Pass AR grade $BaCI_2$ salt through 1 mm sieve and store for use.

Standard stock solution (2000 mg S L^{-1}): Dissolve 1.089 g of oven-dried AR grade potassium sulphate per 100 ml.

Working standard solution (10 mg S L^{-1}): Measure exactly 2.5 ml of the stock solution and dilute to 500 ml.

Barium sulphate seed suspension: Dissolve 18 g of AR grade $BaCl_2$ in 44 ml of hot water and add 0.5 ml of the standard stock solution (given above). Heat the contents to boiling and then cool quickly. Add 4 ml of gum acacia-acetic acid solution to it. Prepare a fresh seed suspension for each estimation.

Dilute nitric acid (approx. 25%): Dilute 250 ml of AR grade conc. HNO_3 to one liter.

Acetic-phosphoric acid: Mix 900 ml of AR grade glacial acetic acid with 300 ml of H_3PO_4 (AR grade).

Procedure

Place 20 g of soil sample in a 250 ml conical flask. Add 100 ml of the monocalcium phosphate extracting solution (500 mg P L^{-1}) and shake for one hour. Filter through Whatman No. 42 filter paper. Measure 10 ml of the clear filtrate into a 25 ml volumetric flask. Add 2.5 ml of 25% HNO_3 and 2 ml of acetic-phosphoric acid. Dilute to about 22 ml, stopper the flask and shake well. Shake the $BaSO_4$ seed suspension and then add 0.5 ml of it and 0.2 g of $BaCl_2$ crystals. Stopper the flask and invert three times and keep. After 10 minutes, invert 10 times and keep. After another 5 minutes, invert 5 times. Allow to stand for 15 minutes and then add 1 ml of gum acacia-acetic acid solution. Make up the volume, invert three times and keep aside for 90 minutes. Invert 10 times and measure the colour intensity at 440 nm (blue filter). Run a blank side by side.

Preparation of standard curve for S

Transfer 2.5, 5.0, 7.5, 10.0, 12.5 and 15.0 ml portions of the working standard solution (10 mg S L^{-1}) into a series of 25 ml volumetric flasks to obtain 25, 50, 75, 100, 125 and 150 µg S. Proceed to develop turbidity as described above for sample aliquots. Read the colour intensity and prepare the curve by plotting readings against sulphur concentration (in µg in the final volume of 25 ml).

Calculation

Av. S in soil mg/100 g soil = (S in aliquot (mg/l) x vol. of extract (ml)}/ (vol. of aliquot (ml) x oven- dry soil (g) x 1000) X 100

1.22 Estimation of Total Sulphur in the Organic Sample After Digestion

Reagents

1. **Gum acacia-acetic acid solution:** Dissolve 5 g of chemically pure gum acacia powder in 500 mL of hot water and filter in the hot condition through Whatman No. 42 filter paper. Cool and dilute to one litre with dilute acetic acid.
2. **Barium chloride:** Pass AR grade $BaCl_2$ salt through a 1 mm sieve and store for use.
3. **Standard stock solution (2000 mg S L^{-1}):** Dissolve 1.089 g of oven dried AR grade potassium sulphate per 100 ml.
4. **Working standard solution (10 mg S L^{-1}):** Measure exactly 2.5 ml of the stock solution and dilute it to 500 ml.

5. **Barium sulphate seed suspension:** Dissolve 18 g of AR grade $BaCl_2$ in 44 ml of hot water and add 0.5 ml of the standard stock solution (given above). Heat the contents to boiling and then cool quickly. Add 4 ml of gum acacia-acetic acid solution to it. Prepare a fresh seed suspension for each estimation every time.
6. **Dilute nitric acid (approx 25%):** Dilute 250 ml of AR grade conc. HNO_3 to one litre.
7. **Acetic-phosphoric acid:** Mix 900 ml of AR grade glacial acetic acid with 300 ml of H_3PO_4 (AR grade).

Procedure

1. Measure 10 ml of the clear filtrate into a 25 mL volumetric flask.
2. Add 2.5 ml of 25% HNO_3 and 2 mL of acetic-phosphoric acid. Dilute to about 22 ml, stopper the flask, and shake well.
3. Shake the $BaSO_4$ seed suspension and then add 0.5 ml of it and 0.2 g of $BaCl_2$ crystals. Stopper the flask and invert three times and keep.
4. After 10 minutes, invert 10 times and keep. After another 5 minutes, invert 5 times.
5. Allow to stand for 15 minutes and then add 1 ml of gum acacia-acetic acid solution.
6. Make up the volume, invert three times and keep aside for 90 minutes.
7. Invert 10 times and measure the colour intensity at 440 nm (blue filter).
8. Run a blank side by side.

Preparation of Standard Curve for S

1. Place 2.5, 5.0, 7.5, 10.0, 12.5 and 15.0 mL portions of the working standard solution (10 mg S L^{-1}) into a series of 25 mL volumetric flasks to obtain 25,50,75,100,125 and 150 µg S.
2. Proceed to develop turbidity as described above for sample aliquots. Where, R stands for the quantity of S in mg as obtained on X-axis against a reading.
3. Read the colour intensity and prepare the curve by plotting readings against sulphur concentration (In µg in the final volume of 25 mL)

Calculation

Total S in Digested Sample in ppm

$$\text{Total S (\%)} = \frac{Factor \times SampleR \times 1000 \times 100 \times 100}{1000 \times 10 \times 1}$$

Where, R stands for the quantity of S in mg as obtained on X-axis against a reading.

1.23 Determination of Ca^{2+} Mg^{2+} in Saturation Extract of Soil by Versenate (EDTA) Method

Principle

Ca + Mg in the solution can be titrated with 0.01 N EDTA using Erichrome black T dye as an indicator at pH 10 in the presence of ammonium chloride and ammonium hydroxide buffer. At the end point colour changes from wine red to blue or green. This titration will estimate both calcium and magnesium. Beyond pH 10 magnesium is not bound strongly to Erichrome black T indicator to give a distinct end point.

Reagents

1. **EDTA or Versenate solution (0.01 N):** EDTA solution (0.01 N): Take 2.0 g of EDTA, dissolve in distilled water and make volume to 1 litre. Titrate with 0.01 N calcium solution by the procedure discussed below and make necessary dilution so that its normality is exactly equal to 0.01 N.
2. **Ammonium chloride + ammonium hydroxide buffer:** dissolved 67.5 g and ammonium chloride in 750 ml of concentrated ammonia and make to 1 litre volume.
3. **Erichorme black T indicator:** Take 100 ml ethanol and dissolve 4.5 g of hydroxyl amine hydrochloride and 0.5 g of the Erichrome black T indictor. Hydroxylamine hydrochloride removes the interference of manganese by keeping it in lower valency state (Mn++), or mix thoroughly 0.5 gram of the indicator with 50 g ammonium chloride.
4. Sodium cyanide solution (2%) or sodium diethyl dithiocarbamate crystals. This is used to remove the interference of copper, cobalt and nickel.

Method

1. Pipette out 10 ml of aliquot (saturation extract) in porcelain dish containing not more than 0.1 me l^{-1} of Ca plus Mg. If the solution has a higher concentration, it should be diluted.
2. Add 0.5 ml of ammonium chloride-ammonium hydroxide buffer. Now add 3-4 drops of Erichrome black T indicator.
3. Titrate this solution with 0.01 N Versenate till the colour changes to bright blue or green and no tinge of wine red colour remains behind.

Observations

S.No.	Volume of aliquot taken (mL)	Burette readings (ml)		The volume of EDTA used (mL)
		Initial	Final	
2				
3				

Calculations

If N_1 and V_1 are normality and volume of aliquot ($Ca^{2+} + Mg^{2+}$) taken and N_2V_2 are the normality and volume of EDTA used, respectively, then $N_1V_1 = N_2V_2$

Or $N_1 = (N_2V_2)/V_1$ = (Vol. of EDTA x Normality of EDTA)/ ml of aliquot taken

Here, N_1, Normality = gram equivalents of Ca^{2+} plus Mg^{2+} present in one litre of aliquot.

Hence,

$Ca^{2+} + Mg^{2+}$ (meL^{-1}) = (Normality of EDTA x Vol. of EDTA) / Volume of aliquot taken*1000

$Ca^{2+} + Mg^{2+}$ (ppm) = $Ca^{2+} + Mg^{2+}$ (meL^{-1}) x equivalent weight (32) = ---------------

Results

The concentration of $Ca^{2+} + Mg^{2+}$ in soil extract = ------------- MeL^{-1}

The concentration of $Ca^{2+} + Mg^{2+}$ in soil extract = ---------------- ppm

1.24 Available Micronutrients (Zn, Cu, Fe, Mn, B and Mo)

Soil tests for micronutrient elements are developing rapidly at present and their quick determination has become possible with the help of atomic absorption spectrophotometer. Myriad studies have been conducted to find a suitable extractant for the simultaneous extraction of available Zn, Cu, Mn and Fe in soils. Lindsay and Norvell (1978) developed a method using DTPA (Diethylene Triamine Penta Acetic Acid) which was found useful for separating soils into deficient and non-deficient categories for Zn, Cu, Mn and Fe in a single extraction by setting the instrument for each of them and using the respective standard curve. This method has been adopted in all the laboratories engaged in the analysis of available Zn, Cu, Mn and Fe in soil in India.

Principle

Chelating agents like DTPA offer great promise for assessing readily available micronutrient cations in soils. These agents combine with free metal ions in solution to form soluble complexes. DTPA offers the most favourable combination of stability constants for the simultaneous complexing of Zn, Cu, Mn and Fe. To avoid excessive dissolution of $CaCO_3$ with the release of occluded micronutrients, the extractant is buffered in a slightly alkaline pH range (7.3) and in part by including soluble Ca^{2+}. Triethanolamine (TEA) is used as buffer because it burns cleanly during atomization and has a pKa = 7.8. At the selected pH of 7.3, three fourth of TEA is protonated and is present as $HTEA^+$. When the extractant is added to the soil, additional Ca^{2+} and some Mg^{2+} enter the solution largely. This raises the concentration of ionic Ca^{2+} by two to three fold and aids in suppressing the dissolution of $CaCO_3$ in calcareous soils with a 2:1 solution to soil ratio. The capacity of DTPA to complex each of the micronutrient cations is 10 times its atomic weight and ranges from 550 to 650 ppm depending on the micronutrient cation. Thus DTPA is present in excess of the micronutrient cations that are normally solubilized during an extraction. This excess reduces the possibility that the extraction of one micronutrient might significantly affect the amount of other micronutrient extracted (Tandon, 1993).

Determination of available zinc

Instruments

Atomic absorption spectrophotometer, electronic balance and mechanical shaker.

Reagents

Dilute HCl : AR grade- Dilute conc. HCl 5 times with double distilled water.

DTPA extractant: Dissolve 1.967 g of AR grade diethyline-triamine-penta acetic acid (DTPA) and 1.470 g of $CaCl_2 \cdot 2H_2O$ (AR grade) in about 25 ml of double distilled water (DDW) by adding 13.3 ml of triethanolamine (TEA), followed by 100 ml more of DDW. Transfer the solution to one liter volumetric flask giving 4 to 5 washings. Just before making up the volume, adjust pH to 7.3 with dilute HCl. This reagent has 0.005M DTPA, 0.1M TEA and 0.01M $CaCl_2 2H_2O$.

Standard stock solution 'A' (100 mg Zn L^{-1}): Weigh exactly 1.0 g of pure Zn metal (AR grade) and dissolve it in minimum volume (about 10 ml) of dil. HCI (1:1) and make the volume to one liter.

Standard solution 'B': Dilute 5 ml of solution A to 100 ml to get solution B containing 50 mg Zn L^{-1}

Standard working solutions : Dilute 0.5, 1.0, 1.5, 2.0, 2.5 and 5.0 ml portions of solution B to 50 ml to get working standards containing 0.5, 1.0, 1.5, 2.0, 2.5 and 5.0 mg Zn L^{-1}. The working standards should be prepared in the medium of the extracting solution after every few days as it cannot be preserved for long.

Procedure

Take 10 g of soil sample in 100 ml conical flask. Add 20 ml of the DTPA extractant and shake for 2 h on a mechanical shaker. Filter through Whatman No. 42 filter paper, discarding first few drops. For quick filtration, Whatman No. 1 filter paper can also be used if the filtrate is clear. Use the filtrate for Zn measurement using AAS. Feed the standard working solutions and prepare a standard curve by plotting AAS readings against Zn concentrations.

Calculation

$$\text{DTPA-extractable Zn (mg 100g}^{-1}) = \frac{A \times Vol\ of\ extractant \times 100}{Wt\ of\ soil \times 1000}$$

$$\text{DTPA-extractable Zn (mg 100g}^{-1}) = A/5 \text{ or } 0.2 \text{ X } A$$

Where,

A stands for the Zn concentration (mg/l) in aliquot as read from X-axis of standard curve against the sample reading.

Determination of Available Copper

Available copper can be determined in the DTPA extract in a similar manner as Zn, using AAS. For this, the standard stock solution can be prepared as given below. Take 1.0 g AR grade copper metal wire or turning and dissolve it in 50 ml of dilute HNO_3 (1:1 with DDW) and finally make up the volume to one liter. This is solution A containing 1000 mg Cu l^{-1}. Prepare solution B containing 50 mg Cu l^{-1} by diluting appropriate volume of solution A. Finally prepare working standards containing 0.25, 0.50, 1.0, 1.5, 2.0 and 2.5 mg Cu L^{-1}.

Determination of Available Iron

Iron in the DTPA extract can also be determined with the help of AAS in the same manner as Zn and Cu described above. However, the working standard solutions of Fe should be prepared for higher concentration as the DTPA – extractable Fe content of the soil is generally more than both Zn and Cu. Thus the Fe standards may be prepared as given below. Prepare standard stock solution (solution A) by dissolving 1.0 g of AR grade Fe metal in about 50 ml of 1:1 diluted HNO_3 and dilute the contents to one liter with DDW. Prepare solution B by diluting 50 ml of solution A to 500 ml to get 100 mg Fe L^{-1}. Finally prepare working standard solutions containing 1.0, 2.0, 3.0, 5.0, and 10.0 mg Fe L^{-1} by diluting appropriate volumes of solution B with the medium of extraction (DTPA).

Determination of Available Manganese

DTPA extractable Mn is also determined following the same technique as adopted for Zn, Cu and Fe. For this, prepare the standard solutions as follows. Weigh 1.583 g of AR grade MnO_2 or 1.0 g of pure Mn metal and dissolve it in 50 ml of dilute HNO_3 (AR grade). Make the volume to 1 liter with DDW to get solution A having Mn concentration of 1000 mg L^{-1}. From solution A, dilute 25 ml to 250 ml with DDW to get solution B having 100 mg Mn l^{-1}. Finally prepare working solutions of 0.5, 1.0, 2.0, 2.5, and 5.0 mg Mn L^{-1} concentration by diluting 0.5, 1.0, 2.0, 2.5, and 5.0 ml portions of solution B to 100 ml.

Analysis of Available Molybdenum

Principle

Ammonium oxalate (pH 3.3) or Grigg's method is considered to be the best one for the determination of available Mo. Further, this extractant is easy to prepare, has sufficient buffering capacity to prevent any material change in the pH of the soil extract and forms stable complexes with Mo ($MoO_3 \cdot C_2H_2O_4$ and $(MoO_3)_2 \cdot C_2H_2O_4$). Molybdate adsorbed on soil colloids and clay is presumably replaced by the oxalate ions. This exchange is made irreversible by the formation of strong Mo-oxalic acid complexes to make a single extraction effective.

Instruments

Colorimeter, hot plate, water bath and electronic balance

Reagents

Potassium iodide solution	:	50% aqueous solution in DDW.
Ascorbic acid solution	:	50% aqueous solution in DDW.
Tartaric acid solution	:	10% aqueous solution in DDW.
Thio urea solution	:	10% aqueous freshly prepared and filtered.

Dithiol solution (Toluene-3, 4-dithiol): To one g of melted dithiol (51°C), 100 ml of 10% NaOH solution is added. The contents are warmed upto 51°C with frequent stirring for 15 minutes on a hot plate. Then 1.8 ml of thioglycolic acid is added and stored in a refrigerator.

Isoamyl Acetate Solution

Ferrous Ammonium Sulphate Solution

Dissolve 63 g of ferrous ammonium sulphate in distilled water and make the volume to 1 liter.

Alcohol for washing of glassware

Extracting solution

To prepare ammonium oxalate (pH 3.3) solution, dissolve 24.9 g AR grade ammonium oxalate and 12.6 g oxalic acid per liter of solution, adjusted to pH 3.3.

Standard solution

Prepare a solution containing 100 μg Mo/ml by dissolving 0.150 g AR grade MoO_3 in 100 ml of 0.1 N NaOH solution rendering it slightly acidic with HCI and make the volume to 1 liter. Take 10 ml of this stock solution in one liter volumetric flask and make the volume to the mark to get a working standard of 1 ug Mo/ml. To make working standards, transfer 0.5, 1.0, 2.0, 3.0 and 4.0 ml of 1 ug Mo/ml solution to series of clean 125 ml separatory funnels.

Standards Curve and Calibration

A widely used colorimetric method for the determination of Mo is by Toluene-3, 4-dithiol, commonly called dithiol. Molybdenum may be determined colorimetrically and this is a widely accepted method. Toluene-3, 4- dithiol yields a slightly soluble dark green complex $(CH_3C_6H_3S_2)_3$ with Mo (VI) in mineral acid medium which can be extracted by an organic solvent. The resulting green coloured complex is used for the colorimetric determination of Mo. The extractable green complex formed between Mo and dithiol is stable for at least 24 hours. Tungston interference is prevented by adding tartaric acid. Tin does not interfere if the absorbance is read at 680 nm. Interference of the ferric iron in the determination is eliminated by reduction with freshly prepared KI and ascorbic acid solution.

50 ml of each working standard along with a Mo-free blank is taken in a separatory funnel (250 ml capacity). Then 0.25 ml of ferrous ammonium sulphate solution and 20 ml of distilled water is added. The excess of KI solution is now added and liberated iodine is cleared by adding ascorbic acid solution drop-wise. After this, 1 ml of tartaric acid and 2ml of thiourea solution are added. After each addition, contents of the separatory funnel are shaken thoroughly. Five drops of dithiol solution are now added and the mixture is allowed to stand for 30 minutes. 10 ml of isoamylacetate are added and the contents of separatory funnels separated in Corning tube/cuvettes. Absorbance of the green complex is measured at 680nm in a spectrophotometer. A standard curve is then drawn by plotting concentration on X-axis and transmittance/absorbance on Y-axis using a semilog graph paper. This graph is used for finding out the concentration of the unknown sample.

Procedure

25 g soil is taken in a 500 ml Corning or pyrex glass conical flask. To this 250 ml ammonium oxalate of pH 3.3 is added. The contents of the flask are shaken over on end to end horizontal shaker for 10 hours and filtered through Whatman filter paper no 50. Then, 200 ml of the filtrate is taken in a 250 ml beaker and evaporated to dryness on a water bath. The contents of the beaker are heated at high temperature (500 °C) in a furnace for 50 hours to destroy organic matter and oxalates.

Contents of the beaker are then digested with 5 ml of diacid (HNO_3:HClO4, 4:1) after keeping it overnight, then with 10 ml of 4N H_2SO_4 and H_2O_2 every time taking to dryness. Then 10 ml of 0.1 N HCl is added to the beaker and filter paper washed with another 10 ml of 0.1 N HCl and redistilled water until the filtrate volume obtained is 100 ml. The filtrate is stored in corked plastic reagent bottles for estimation of available Mo by the procedure described above under standard curve.

Calculation

$$\text{Available Mo in soil (mg per 100 g soil)} = \frac{A \times 250 \times 100}{100 \times 25 \times 1000}$$

Or

Available Mo is soil (mg per 100 g soil) = A/200

Where,

A stands for Mo conc. in mg as obtained from x-axis against sample reading

Precautions

Keep separatory funnels on a revolving type separatory funnel stand. Wash all glassware with alcohol before use. Dry separatory funnels, before use. Separate out aqueous phase first and take out the organic phase (green coloured complex). While estimating Mo, take out dithiol solution from refrigerator and keep the bottle in an ice box. Don't store dithiol solution for more than one week. Remove stopper from separating funnel before taking out the two phases.

Analysis for available Boron

Out of several methods devised to assess the level of available B in soil, the hot water soluble B method of Berger and Truog (1939) has been most widely accepted. Recent description of this method includes some changes (Keren and Bingham, 1985 and Sippola and Ervco, 1987), but the basic procedure remains the same. Throughout boron analysis, use of borosilicate glassware should be avoided even for storage of chemicals. Plastic containers or Corning/Pyrex glassware should be used.

Instruments

Colorimeter or spectrophotometer, refrigerator and polypropylene test tubes of 10 ml capacity.

Reagents

Buffer solution: Dissolve 250 g of ammonium acetate and 15 g of EDTA – disodium salt in 400 ml of distilled water and slowly add 125 ml of glacial acetic acid and mix.

Azomethine-H reagent: Dissolve 0.45 g of azomethine-H in 100 ml of 1% L-ascorbic acid solution. Fresh reagent should be prepared every week and stored in a refrigerator.

Standard Solution: To prepare the standard stock solution, dissolve 0.570 g boric acid (H_3BO_3) AR grade in a liter of distilled water to obtain a stock solution of

100 µg B/ml. Take 5 ml of the stock solution in a 100 ml volumetric flask and dilute to the mark. This solution contains 5 µg B/ml.

Working Standards

To a series of 25 ml volumetric flasks, add 0.25, 0.5, 1, 2 and 4 ml of 5 µg B/ml solution and make the final volume to 25 ml.

Volume of stock solution taken = 0, 0.25, 0.50, 1.0, 2.0, 4.0 ml

Concentration of B in solution = 0, 0.05, 0.10, 0.20, 0.40, 0.80 µg B/ml.

Preparation of Standard Curve

To each volumetric flask containing working standards, add 2 ml of buffer solution and mix, add 2 ml of Azomethine-H reagent solution, stir thoroughly and allow to stand at room temperature for 30 minutes. Make the volume to 25 ml with distilled water and measure absorbance at 420 nm on Spectronic 20 or any other spectrophotometer.

Over a concentration range of 0.5 to 10 µg B/ml, Azomethine-H solution forms a stable complex with H_3BO_3 at pH 5.1. Maximum absorbance occurs at 420 nm with little or no interference from a wide variety of salts. Plot the graph of concentration vs absorbance on a semilog paper, plotting concentration on X axis and absorbance on Y axis. Use distilled water for the 0.0 µg B/ml standard solution.

Extraction and Determination of Boron

Take 25 g soil sample, 50 ml of water and about 0.5g of activated charcoal and boil for 5 minutes in a quartz flask and filter immediately thorough Whatman filter paper No. 42. 5 ml of the extract is taken in a 25 ml volumetric flask and 4 ml of buffer solution and 4 ml of azomethine-H reagent solution is added. The colour is allowed to develop for 1 hour, and the volume is made to the mark. Intensity of colour is measured spectrophotometrically at 420 nm, and the B concentration is read off from standard curve described above.

Calculation

Boron (µg/g or ppm) in soil sample = concentration of B in analysed sample (reading from the standard curve x 10).

1.25 Favorable Range of Soil Parameters For Aquaculture

i. **Soil texture** : sandy clay /sandy clay loam /clay loam

ii. **Soil colour** : blackish brown

iii. **pH** : 6.0 to .8.0

iv. **Water retention capacity** : 40% and above

v. **Sand**: 40%

vi. **Silt** : 30%

vii. **Clay** : 30%

viii. **Total nitrogen** : 50 mg/100 gm (0.05%) of soil sample and above

ix. **Phosphate** : 6 mg/ gm of soil sample and above

x. **Potassium** : 25 mg /100 gm of soil sample and above

xi. **Organic carbon** : 0.5% and above

xii. **Electrical conductivity** : less than 16 millimhos/cm

SOIL MICROBIOLOGICAL PARAMETERS

1.26 Estimation of Microbial Biomass Carbon in Soil

It was estimated by chloroform fumigation extraction method (Brookes *et al.*, 1985).

Principle

Overnight fumigation of chloroform is done to kill all the organisms in soil samples; after which the amount of the organic C in the sample can be measured by fumigation- extraction method.

Fumigation Extraction Method

The microbial biomass constituents released by $CHCl_3$ fumigation treatment can be extracted directly through chemical extractants and the readily oxidisable C contained in the extract can be measured through standard chemical procedures.

Reagents

1. Ethanol free chloroform
2. Conc. H_2SO_4
3. 0.5M K_2SO_4: 43.563g K_2SO_4 was dissolved in distilled water and diluted with 500 ml.
4. 0.2 N $K_2Cr_2O_7$: 0.9808 g $K_2Cr_2O_7$ was dissolved in 100 ml of distilled water.
5. Orthophosphoric acid
6. 0.005 N Ferrous Ammonium Sulphate (FAS): 3.92 g of ferrous ammonium sulphate was dissolved in 0.15 ml conc. H_2SO_4 and then diluted to 2 liters

by distilled water.

Ferroin/ Diphenylamine Indictor

Procedure

1. 10g of soil (three sets of each soil sample) was weighed and kept in to 100 ml beakers each.
2. 8 ml of distilled water was added to both beakers and were incubated for seven days at 25^{0}C in an incubator
3. 20 ml of chloroform was taken in a separatory funnel. It was washed twice with conc. H_2SO_4 (with half of the volume of chloroform) and the acid (bottom phase) was discarded. It was washed twice with the same volume of distilled water similarly to make the chloroform free of ethanol and the bottom whitish phase was collected.
4. One set of soil samples was taken in a crucible and fumigated with ethanol free chloroform in a vacuum desiccator. For this purpose 20 ml ethanol free chloroform was taken in a petri dish and was placed inside the desiccator the on bottom portion inside the vacuum desiccators. It was attached to the vacuum pump and the air was evacuated until the chloroform starts boiling to saturate the desiccators with chloroform fumes. Then the vacuum desiccator was kept in dark room for overnight.
5. Next day the vacuum was released and chloroform was removed from the desiccator.
6. 10 g each of fumigated and non-fumigated soil samples were weighed in 150 ml conical flask and 40 ml of K_2SO_4 (0.5 M) was added to each flask. Samples were shaken for 30 minutes on a rotary shaker.
7. Both the samples were filtered with Whatman no. 42 filter paper, labelled and frozen until digestion.
8. 10 ml of the filtrate was taken in 100 ml conical flask and 2 M of $K_2Cr_2O_7$ (0.2 N) and 10 ml of con. H_2SO_4 was added to it. The contents of the flask were allowed to cool for half an hour, then 5 ml orthophosphoric acid was added along with 200 ml distilled water. Minimum two blanks were also run with 10 ml distilled water.
9. Few drops of diphenylamine indicator were added and titrated against ferrous ammonium sulphate (0.2 M) till the colour changed from violet to green.

Microbial carbon(ppm) = (Fumigated C –Non-fumigated C)/ 0.44

Microbial biomass estimation

1.27 Total Plate Count (TPC)

The plate count method allows the enumeration of viable cells. The plate count method depends on bacteria forming a colony on a nutrient medium. The colony becomes distinct and is observable to the naked eye, now the number of colonies formed on a plate can be counted.

Media

☆ ***Nutrient agar granulated (Bacteria)***

7 g of nutrient agar + 1.5 g of agar powder 250 mL D.W and autoclave

☆ ***Rose Bengal Agar base (yeast/fungus)***

7.88 RB G of RB + 1.5 g of agar powder 250 ml D.W and autoclave

☆ ***Kenknight & Munaier's Medium (Actinomycetes)***

1g of KM + 1.5 g of agar powder 250 mL D.W and autoclave

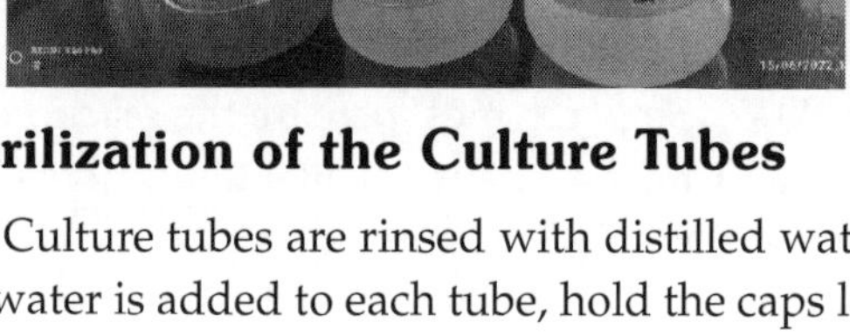

Sterilization of the Culture Tubes

Culture tubes are rinsed with distilled water and dried. After drying, 9 ml of tap water is added to each tube, hold the caps loosely, and autoclave @ 121°C.

Procedures

- ☆ Sterilize culture tubes and media and keep them in laminar flow
- ☆ The samples (serial dilution) are labelled as 10^{-1} , 10^{-2} , 10^{-3} ,10^{-4}
- ☆ Pour the media into a Petri plate and let it solidify
- ☆ 1g of soil is taken and kept in 10^{-1} dilution, followed by serial dilution
- ☆ 0.1ml from 10^{-1} is pipetted in solidified Petri plate
- ☆ The spreader is dipped in alcohol and sterilized in flame, spread the sample carefully and adequately
- ☆ Continue for 10^{-1} , 10^{-2} , 10^{-3} , and 10^{-4} for all three media.
- ☆ Mark the plate, the sample name, no. of replication, specific medium and date
- ☆ Leave the plate for incubation at 28°C
- ☆ The sample are kept for 72 hrs and colonies are counted
- ☆ No. of CFU/g of sediment = No. of colonies / dilution *vol. of innoculum in plate

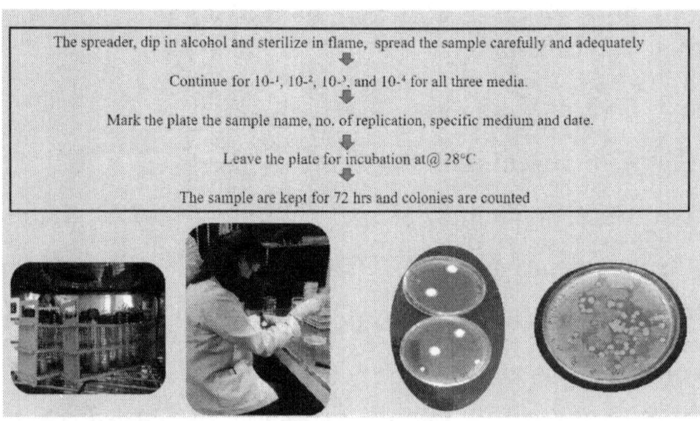

1.28 Estimation of Soil Enzymes

Soil enzymes are proteins that act with specific substrates to catalyse soil biochemical reactions necessary for microbial life functions. Enzymes increase the reaction rate at which organic matter decomposes and releases nutrients into soil environment. Soil enzymes can be constitutive (constantly produced) or induced and can originate from both living and dead micro-organisms soil animals and plant roots (Tabatabai, 1994). Enzyme activity can serve as a sensitive indicator of ecological or microclimatic change (Kandeler, 2015). Soil enzymes increase the reaction rate at which plant residues decompose and release plants available nutrients. The substance acted upon by the soil Enzyme is called the substrate.

Importance of Soil Enzyme Study

1. Soil enzymes play an important role in organic matter decomposition and nutrient recycling.
2. Enzymes respond to soil management changes long before other soil quality indicator changes are detectable.
3. Some soil enzymes only facilitate the breakdown of the organic matter (i.e hydrolase, glucosidase), while others are involved in the nutrient mineralization (e.g. amidase, urease, phosphatase and sulfatase).
4. Soil enzymes as sensitive indicators of ecological changes.
5. Mediator and catalyst of important soil functions that include like decomposition of organic inputs, transformation native organic matters, release of inorganic nutrient for plant growth etc

Dehydrogenase Activity (DHA) (Casida,et al.,1964)

Respiration is an index of biological activity. Monitoring the activity of respiratory enzymes gives a measure of biological activity of soil. Determination of DHA is done by adding alternative electron acceptors to soil samples, as dehydrogenase oxidises the substrate. TTC is preferred oxidant because it formed water- insoluble coloured formazans which can be measured spectrophotometrically.

Reagents Required

- ☆ 2,3.5- Triphenyl tetrazolium chloride (TTC) solution (3%):
- ☆ Methanol (AR grade)
- ☆ Glucose solution
- ☆ Standard Triphenyl formazan(TPF, 100ppm)

Procedure

- ☆ 2 g of soil is taken into culture tubes
- ☆ Add 0.6 ml of distilled water, incubate at 30°C for 24 hrs
- ☆ Add 0.1 ml of TTC(3%) except blank to each tube, incubate at 30°C for 24 hrs Add 10 ml of methanol and for blank, TTC(3%) is added, and 10 ml of methanol is immediately added
- ☆ Caps are closed tightly and shaken well. Wait for 10 minute
- ☆ The supernatant is collected in a 25 ml volumetric flask using Whatman filter no.42
- ☆ Repeat the extraction with methanol and make up the volume to 25 ml with methanol

☆ The intensity of red to pink colour developed is read in a spectrophotometer at 485 nm, and readings are recorded.

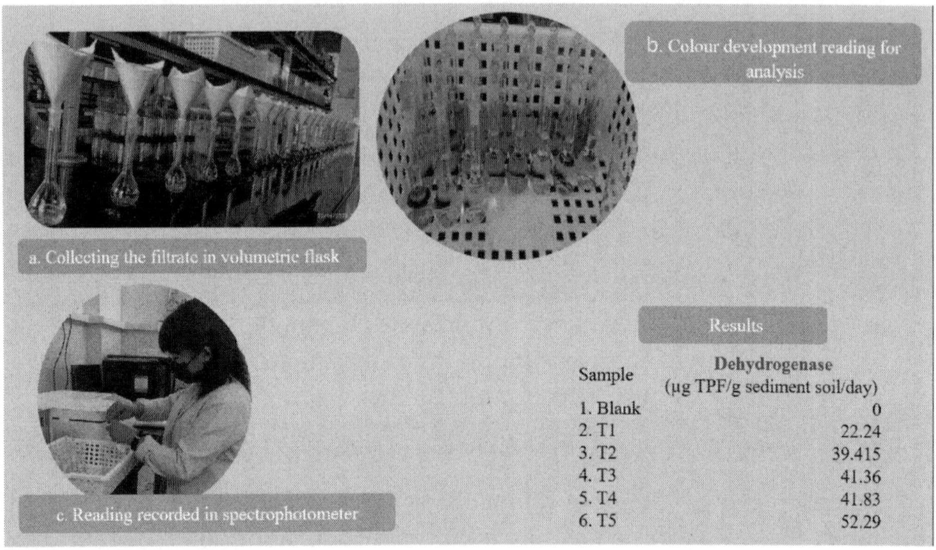

Sample	Dehydrogenase (µg TPF/g sediment soil/day)
1. Blank	0
2. T1	22.24
3. T2	39.415
4. T3	41.36
5. T4	41.83
6. T5	52.29

Fluorescein Diacetate (FDA) Hydrolysis (Adam and Duncan, 2001)

Fluorescein diacetate (FDA) hydrolysis assay is used to measure the **enzyme activity of microbes** in a sample. A bright **yellow-green glow** is produced and is strongest when enzymatic activity is greatest.

Reagents :

1. 60mM potassium phosphate buffer pH 7.6
2. 2:1 Chloroform/methanol
3. 1000 ug Fluorescein/ml stock solution:

Procedure

☆ 2g of soil is added to a 50 ml centrifuge tube, add 0.6 ml of distilled water, incubate for 24 hrs in a 30°C incubator

☆ 0.2 ml of FDA solution is added to all the flasks except blank. This solution initiates the reaction

☆ Tubes are capped and shook properly by hand and placed in an orbital shaker at 30°C for 20 mins.

☆ Add 15 ml of 2:1 Chloroform and methanol (v/v). Wait for 10 minutes.

☆ The supernatant is filtered using Whatman no. 42

☆ Colour development is measured at 490 nm on a spectrophotometer.

Phosphatase Enzyme: (Acid and Alkaline) Tabatabai & Bremner, 1969

Procedure-

- ☆ Place 1g soil in 50ml conical flask
- ☆ Add 0.2 ml toluene
- ☆ Add 4 ml MUB (Modified Universal Buffer) (pH 6.5 for acidic and pH 11 for alkaline phosphatase)
- ☆ Add 1 ml PNP(p-nitrophenyl phosphatase) (Not in control)
- ☆ Closed with the stopper and incubate at 37°C for 1 hour
- ☆ Remove the stopper
- ☆ Add 1 ml 0.5 M CaCl2
- ☆ Add 4 ml NaOH
- ☆ Add PNP in control. Swirl/mix
- ☆ Filter through Whatman No 42 filter paper
- ☆ Measure the colour intensity at 410 nm

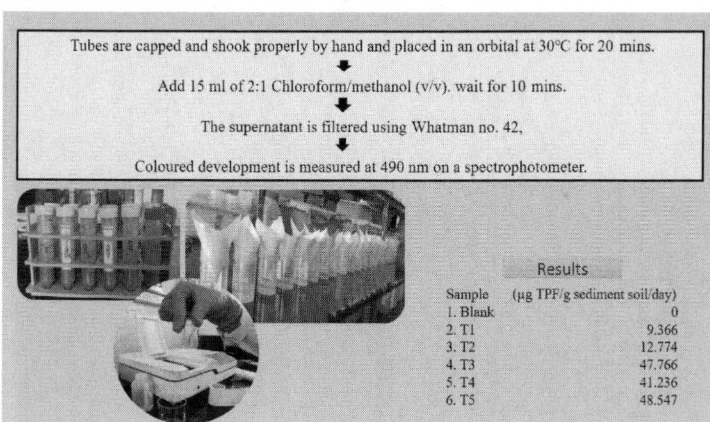

Chapter 2

WATER TESTING

Water is the most valuable universal solvent and all its naturally occurring sources are bound to contain variable quantities of dissolved solids and gases and sometimes colloidal organic and inorganic materials are also present as well. In fish culture, water quality is determined to know its total nutrient load and to judge the suitability of water for the survival and growth of fish and contribution to nutrient supply. Water analysis is carried out to assess the optimum and harmful limits of various parameters for survival and general health of fish. Physico-chemical properties of a water body play a major role in its productivity process and growth of aquatic organisms under culture. Selection of aquaculture species is basically made on the basis of prevailing physico-chemical properties of the environment in a particular eco-system. These parameters vary on the basis of geographical location. The analysis of these parameters is important in order to diagnose the constraints or opportunities in the aquatic system and to explore management options for optimum economic return.

2.1 Collection of water sample

Sampling means to collect a portion of water small enough in volume to be conveniently transported to and handled in the laboratory, and representative for the bulk. Reliable results will be obtained if the sampling procedure is executed satisfactorily. This means proper preparation in the laboratory before sampling, accurate field work and finally, proper handling and cleaning when entering the laboratory.

Rinse the sample bottles in the laboratory with HCl (2M), distilled water (at least three times) and dry them. On the sample location, rinse again three times with sample water before the collection of sample. Always fill a sample bottle completely and fill the screw cap too. Execute this work with analytical precision and cleanliness. Don't touch the inner part of the sample bottle, screw cap or even the water sample itself. Time interval between collection and analysis of the sample should be short (not more than 72 hours) in order to get reliable results. If a delay is inevitable, water sample must be protected against bacterial growth either by adding 2-3 drops of pure toluene or by keeping it under refrigeration (4°C) for the time required. Determination of temperature, pH and dissolved gases should always be carried out in the field, as otherwise changes are almost inevitable by the time the samples reach the laboratory. Fortnightly sampling is desirable. As a general rule, results of chemical analysis of water are expressed as milligrams per liter (mg/l) or mill equivalents (me/l) per liter. Then me/l is obtained by dividing the concentration expressed in mg/l by the equivalent weights of the ions or the compound. Analysis of condition and constituents of water that are of primary importance is discussed below.

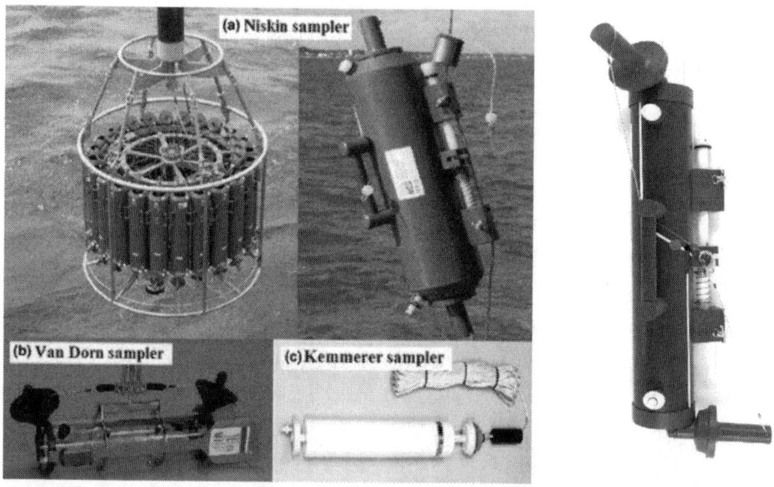

The commonly used samplers for deep seawater sampling. (a) Niskin sampler (http://www.godac.jamstec.go.jp/darwin/instrument/mirai/e), (b) Van Dorn sampler (http://www.kc-denmark.dk/products/water-sampler/van-dorn-water-sampler.aspx) and (c) Kemmerer sampler (http://www.rickly.com/as/kemmerer.htm).

2.2 Temperature

Temperature plays an important role in limnological studies. It is usually determined by a centigrade thermometer of 0.1°C or of 0.01 °C marking when more accuracy is desired (Jhingran et al, 1988). Surface water temperatures may be determined by collecting the water in a large plastic container and dipping the thermometer directly into it, keeping it steady for about a minute and then noting the temperature. Temperature for water samples is to be determined immediately after collection. As the solubility of oxygen is dependent on temperature, it should be noted for the dissolved oxygen samples before fixing.

Subsurface temperature may be determined accurately by a reversing thermometer. This thermometer can be placed in any desired depth from a boat and by means of a trigger arrangement, the thermometer is reversed at that point when the mercury column becomes fixed. It can then be taken up, temperature is noted and the thermometer is reset for another observation. Indicate the temperature after effecting necessary corrections to the reading.

The mean temperature of the water source is derived from the vertical temperature records

- ☆ by adding the temperatures limiting the upper (t_1) and lower (t_2) limits of each submerged stratum and dividing by 2 as (t_1+ t_2)/2 and
- ☆ by multiplying the above (a) by the percent by volume of each substratum in the volume of water source, being designated here as P and (b) summiting all terms (t_1+ t_2) P corresponding to each stratum which constitutes the mean temperature

2.3. pH

There are various methods for determining water pH. These may be divided into electrometric and colorimetric methods.

Electrometric Estimation

Principle

pH expresses the intensity of the acid or alkaline reaction of a solution in terms of the hydrogen ion concentration but is not a measure of the total concentration of acid or alkali present. It is the logarithm of the reciprocal of the hydrogen ion concentration in moles per litre. $pH = \log_{10} (H^+)$. Water containing no acid or alkali has a pH value of 7.0 which is termed the neutral pH value. pH values below 7.0 indicate acidity, pH value above 7.0 indicates alkalinity. Most natural waters have pH values between 5.5 and 8.6.

Equipment: pH meter

Procedure

The pH meter is allowed to warm up for 10 minutes and then it is adjusted to the known pH of buffer solutions having pH 4.0, 7.0 and 9.2. Then 50 ml water sample is taken in a 100 ml clean beaker and electrodes of the pH meter are immersed in it. The pH is indicated on the dial and is recorded. It should be ensured that the calomel electrode is clean and filled with saturated KCl solution before making the measurement.

Colorimetric Estimation

Colorimetric methods that are more widely used can be taken up for clear water with sufficient accuracy but difficult to use for highly turbid or coloured water. Such waters are often made sufficiently clear by stirring them with a little neutral barium sulphate, allowing it to settle and then decanting the clear liquid. At times, however, they present considerable difficulties and the results suffer from inaccuracy (Jhingran et al. 1988).

Principle

The principle of colorimetric estimation of pH is to develop colour in the sample with an indicator dye and to compare this with colour of glass discs, colour charts or coloured buffer solutions.

Place 10 ml of the sample in a small clear glass tube and add 0.5 ml of indicator. To know which of the above indicators is to be used, a preparatory test with a universal indicator may be done which gives a very approximate value of the pH; otherwise phenol red would be used first and then if necessary indicators for higher or lower ranges are used. As the pH of inland waters generally varies between 6.0 and 9.6, the use of three indicators viz., Bromo-thymol blue, Phenol red and Thymol blue may be sufficient. After adding the indicator to the sample, it is stirred gently and the colour developed is matched against colour discs in a comparator or standard colour charts, in the absence of which buffer solutions of different pH values may be prepared, treated similarly with indicators and kept in sealed tubes for comparison.

The indicators used with range of pH are given below:

Indicator	pH range
Bromo phenol blue	3.0-4.6
Bromo cresol green	3.8-5.4
Bromo cresol purple	5.2-6.8
Bromo thymol blue	6.0-7.6
Phenol red	6.8-8.4
Cresol red	7.2-8.8
Thymol blue	8.0-9.6

2.4 Electrical Conductivity (Total soluble salts)

Principle

Amount of total soluble salts in a sample is generally expressed in terms of the sample's electrical conductivity (EC). A simple Wheatstone bridge circuit is used to measure EC by null method. It is expressed as dsm^{-1} that is equivalent to the olden unit millimhos/cm. The temperature is kept at 25 °C.

Equipment

Conductivity meter or solubridge

Reagent

1 N KCl solution

Procedure

Take the water sample in a clean 100 ml beaker. Rinse the conductivity cell with distilled water and then with the sample. Adjust the temperature and cell constant corrections on the conductivity meter. Connect conductivity cell to meter and dip in the sample. Pass the current and adjust the current by rotating the dial in such a way that the width of shadow in cathode ray is maximum. Then read the conductivity value in dsm^{-1}. Direct reading is obtained in digital type of meters.

2.5 Transparency (Light Penetration)

Light quality, intensity and photoperiod are all important to phytoplankton growth and may exert considerable influence on fish as well. Light penetration basically depends upon the available intensity of the incident light that varies with the geographical location of the pond. Shade of the surrounding vegetation affects the incidence of light on the pond. Light penetration in water is also affected by the prevailing turbidity. The photoperiod requirements of various aquaculture species and their various stages in the life cycle process vary considerably. The onset of sexual development is greatly influenced by photoperiod. Alterations in photoperiod help in maturation and spawning of aquaculture species. For better primary productivity, an illumination of 1000-8000 lux is necessary. Secchi disc transparency between 30 and 40 cm indicates optimum primary productivity of a pond.

Instrument

Secchi Disc: Secchi disc is commonly used to measure the light penetration. It is a 20 cm diameter circular metal plate. The upper surface is divided in to four equal quadrants. Two quadrants located diagonally are painted black and other two are painted white. A staple fixed at the center of upper surface provides attachment for the graduated rope. On the opposite side of the staple, a weight is fixed at the lower surface that facilitates the sinking of the disc in proper position. Lower surface of the disc is painted with black colour in order to eliminate reflection of light.

Procedure

Dip the Secchi disc vertically in the water by holding the graduated rope at the water surface. Note the reading when Secchi disc disappears and reappears while lifting the rope. The average of the two readings gives the limit of visibility that is expressed in centimeters (cm).

Calculation

$$\text{Transparency} = (A+B)/2$$

where,

A = depth at which secchi disc disappears

B = depth at which secchi disc reappears on lifting the rope

Note: Keep the Secchi disc surface clean and bright. Do frequent painting of the disc. Observations should be made during mid-day when sky is clear and sun is overhead. Use Secchi disc under a sun shade. Avoid taking reading in the presence of heavy waves or ripples.

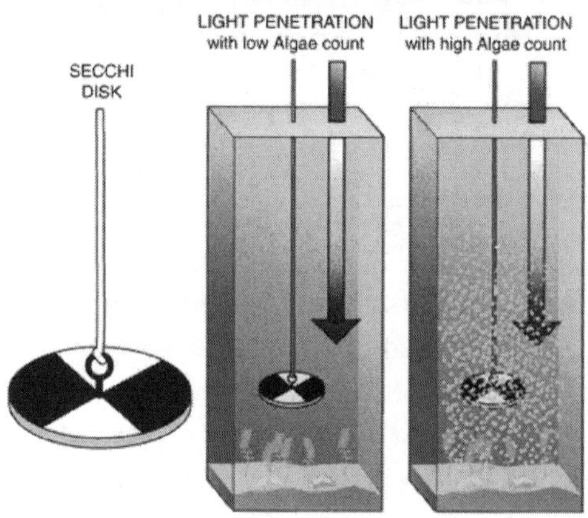

2.6 Turbidity

Turbidity is the decreased ability of water to transmit light caused by suspended particulate matter and phytoplankton. Turbidity of pond water has both advantages and disadvantages. The higher the concentration of suspended solid, there is increase the turbidity. If turbidity becomes too high due to suspended inorganic solids, primary productivity may be reduced because of restricted penetration of sunlight. This can be an advantage in reducing the growth of filamentous algae and aquatic macrophages, but it can be disadvantageous if phytoplankton are required to provide food for young aquatic organisms or herbivores.

Equipment

US Geological survey turbidity rod is used to measure turbidity. It is a 6 mm thick, 17 mm wide and 1 m long made of non-rust metal rod marked from 3000 to 10. A piece of straight, bright platinum wire of 1 mm diameter and 25 mm length is mounted in a detachable screw eye that is fixed at the end of the scale. An eye point is marked to view the platinum wire.

Procedure

Hold the turbidity rod straight and slowly dip vertically into the water by observing the platinum wire from the eye point mark. When the platinum wire fades from view, record the surface level of water on the graduated rod. This value is the measurement of turbidity of sample water in parts per million (ppm). Waters having turbidity exceeding 500 ppm are diluted with clear water and then read by turbidity rod. Compute the turbidity of original sample from the dilution data.

Note: Make all observations in the open air, preferably during mid-day but not in direct sunlight. Keep the platinum wire clean and bright. Avoid taking readings in the presence of waves, ripples or other disturbances.

2.7 Dissolved Oxygen

In contrast to atmosphere, oxygen becomes a limiting factor for aquatic organisms because the saturation concentration of oxygen in water is governed by temperature and salinity. The lower the temperature, the greater is the oxygen retaining capacity of water. In aquaculture, the oxygen consuming (respiration) and oxygen producing (photosynthesis) processes should be so maintained that it always remains within congenial limits. The D.O. concentration in the range of 5.0 mg/l to 10.0 mg/l is desirable for better survival and growth of the fish/prawns. After collection, the sample should be fixed immediately and taken to laboratory for analysis or analyzed on the spot.

Reagents

Manganous sulphate (Winkler's-A) solution (Dissolve 480 g of $MnSO_4$. $4H_2O$ or 400 g of $MnSO_4$. $2H_2O$ or 364 g of $MnSO_4$. H_2O in distilled water and make up the volume to one liter)

Alkaline potassium iodide (Winkler's-B) (Dissolve 500 g of NaOH, 150 g of KI and 20g sodium azide in distilled water and make up the volume to one liter).

Standard sodium thiosulphate (0.025 N or N/40) solution (Dissolve 6.205g of $Na_2S_2O_3$ $5H_2O$ in distilled water and make up the volume to 1 liter).

Starch indicator Make thin paste of about 1g of starch in cold water. Pour 200ml of boiling water in it and stir. Add 0.1g of salicylic acid or 0.5g of sodium chloride or 0.5ml of formalin as preservator.

Procedure

Collect the sample in 100 ml glass stoppered sample bottle. Avoid contact of the sample with air. The bottle should be completely filled up to the brim. Immediately after collection, add 1 ml of manganous sulphate solution by means of a pipette, dipping the end of the pipette at the bottom and gradually pulling it near the surface of the water. Add 1 ml of alkaline potassium iodide solution in a similar manner. Insert the stopper and mix by inverting the bottle several times with jerks. Allow the precipitate to settle half way and mix again. Allow the precipitate to settle. If the precipitate is whitish in colour, oxygen is very poor, brown colour indicates high dissolved oxygen. Add 1ml of conc. Sulphuric acid. Insert the stopper after the addition of the acid and mix as before. Allow the solution to stand for at least 5 minutes. Withdraw 50 ml. of the solution to a conical flask and add 0.025 N sodium thiosulphate drop by drop from a burette until the yellow colour turns to straw colour. Add 4-6 drops of starch solution and continue the addition of thiosulphate until the blue colour turns to straw colour. Record the volume of thiosulphate consumed.

Calculation

$$\text{Dissolved Oxygen} = \frac{\text{ml of 0.025 N sodium thiosulphate x 1000 x 8}}{50x40}$$

2.8 Dissolved Free Carbon Dioxide

Free carbon dioxide is highly soluble in water. It enters the water from the atmosphere at the air-water interface and naturally present in solution as a by-product of metabolism. Water supporting good fish population normally should contain less than 3mg/L CO_2.

Reagents

Phenolphthalein indicator (Dissolve 5g of phenolphthalein in 50% alcohol and make up the volume to 1 liter).

N/44 NaOH (Dissolve 0.909g of NaOH pellets in distilled water and make up the volume to 1 liter).

Procedure

Take 50 ml of the water sample in a Nessler's tube and add two drops of phenolphthalein indicator. If the water turns pink, there is no free carbon dioxide. If the water remains colourless, add N/44 NaOH drop-wise through a 1 ml graduated pipettle with gentle stirring using glass rod till the colour turns pink. Record ml of N/44 NaOH consumed during the titration.

Calculation

Dissolved free CO^2 (mg/l) =

$$\frac{Vol.\ of\ N/44\ NaOH\ consumed \times 1000 \times 44}{50 \times 40}$$

2.9 Alkalinity

Alkalinity is the sum of negative ions viz. carbonate, bicarbonate and hydroxyl ions reacting to neutralize hydrogen ions when an acid is added to water. Alkalinity is normally expressed in mg / l. The alkalinity of freshwater is lower and more variable than sea water. The amount of CO_3^{2-} and OH^- is determined by titration with acid to pH 8.3, the end point being detected with phenolphthalein. The amount of HCO_3^- is determined by further titration with acid to an end point pH between 4.2 and 5.4, with methyl orange or mixed indicator as end point indicator.

Reagents

N/50 H_2SO_4 (Prepare 1N H_2SO_4 by adding 1 ml conc. H_2SO_4 in 35 ml distilled water. Add 1ml of this 1 N H_2SO_4 solution to a 49 ml distilled water).

Phenolpthalein indicator. (Dissolve 0.5g of phenolphthalein in 50 ml of 95% ethanol and add 50 ml of water. Add dilute (0.05 M) CO_2 free NaOH solution dropwise, until the solution turns faintly pink).

Methyl orange indicator. Dissolve 0.05 g of methyl orange in about 100 ml of water. This indicator is suitable for equivalence points below pH 4.6.

Procedure

Mix 100 ml of the sample with 2 drops of phenolphthalein indicator in a conical flask. If the solution turns pink, titrate with N/50 H_2SO_4 until the colour practically disappears. Record the volume of acid used. Add 3 drops of methyl orange indicator in the conical flask and titrate with N/50 H_2SO_4. The endpoint is orange. The endpoint can be best judged by comparing it with the blank.

Calculation

Total alkalinity (mg/l) = (Vol. of H_2SO_4 (N/50) consumed x 1000)/ml of sample

2.10 Salinity

Principle

The term salinity refers to the total concentration of all dissolved ions in water. In a neutral of slightly alkaline solution, potassium chromate can indicate the end point of silver nitrate titration of chloride. Silver chloride (AgCl) is precipitated quantitatively before red silver chromate is formed.

Reagents

Silver nitrate solution of 0.28 N (Dissolve 48.5g of $AgNO_3$ in one litre of distilled water)

Potassium chromate indicator (Dissolve 63 g of K_2CrO_4 in 100 ml distilled water. Add a few drops of 0.28 N $AgNO_3$ until a definite red precipitate persists)

Procedure

Pipette out 10 ml of sample in a flask and add 2 or 3 drops of K_2CrO_4 indicator; colour of sample turns yellow. Titrate it against 0.28 N $AgNO_3$ drop by drop so that the solution becomes reddish brown that does not disappear on shaking. This gives the end point.

Calculation

Salinity (ppt) = Burette reading + 0.0355 (correction factor)

The above relationship between salinity and chlorinity is empirical, developed from the water density at 20 °C.

1.11 Hardness

Hardness is the measure of total concentration of divalent metallic cations like calcium, magnesium, strontium and expressed in mg/l.

Reagents

Standard EDTA solution Dissolve 4 g of sodium salt of EDTA and 0.1 g of $MgCI_2\,6H_2O$ in 800 ml of distilled water, and dilute the solution to a volume of 1 liter.

Ammonia buffer Dissolve 6.75 g of NH_4Cl in 57 ml of liquid ammonia and dilute to 100 ml distilled water.

Eriochrome black – T indicator Dissolve 0.1 g of E Black-T in 20 ml of ethyl alcohol. Prepare a fresh solution at every 3 weeks period.

Procedure

Take 100ml of the sample water in a conical flask. Add 1 ml ammonia buffer and 3 drops of EBT indicator. Titrate with standard EDTA solution till the colour from the wine red to blue. Record the volume of EDTA solution used.

Calculation

Total hardness (mg/l) = (vol. of EDTA solution used)/ vol. of sample × 100

2.12 Dissolved Solids

Materials Required

Porcelain dish, silica crucible, Corning or Pyrex beaker (500ml capacity), funnel, oven with thermostatic control, desiccator, Whatman filter paper no. 44 and electronic balance.

Procedure

For total solids: Place the sample in a weighed dry dish/crucible/corning or pyrex beaker. Evaporate to dryness in an oven at 103 °C to 105 °C. Cool the container in a desiccator. Weigh and note the weight.

$$\text{Total solids (mg/l)} = \frac{\text{mg residue} \times 1000}{\text{ml of sample}}$$

For total suspended solids: Filter the sample water through Whatman No. 44 filter paper. Take a suitable quantity in a dry dish / crucible/ corning or pyrex beaker. Evaporate to dryness in an oven at 103 °C to 105 °C. Cool the container in a desiccator. Weigh and note the weight.

$$\text{Total dissolved solids (mg/l)} = \frac{\text{mg residue} \times 1000}{\text{ml of sample}}$$

Calculation

Total suspended solids mg/l. = Total solids – total dissolved solids

2.13 Biochemical Oxygen Demand

The BOD test is widely used to determine

- ☆ The degree of pollution in lakes, ponds, rivers, etc. at any time and their self-purification capacity.
- ☆ Pollution load of water and
- ☆ Efficiency of waste treatment plant.

The biochemical oxygen demand (BOD), is the amount of oxygen taken up by micro-organisms that decompose organic waste matter in water. It is, therefore, used as a measure of the amount of certain types of organic pollutants in water. Standard BOD determination is done by incubating samples for 5 days at 20°C and is denoted by BOD_5 but for fish ponds, a short term BOD_1 procedure could be adopted by incubating the water sample in the pond itself for 24 hours and determining the decrease in its dissolved oxygen concentration (BOD, at prevailing pond temperature). The BOD is a measure of the biologically decomposable organic matter in a sample under aerobic conditions.

Reagents

Phosphate buffer (Dissolve 8.5 g of KH_2PO_4 21.75 g of K_2HPO_4 and 33.4 g $Na_2HPO_4.7H_2O$ in 500 ml of distilled water and make up the volume to one liter).

Magnesium sulphate solution (Dissolve 22.5 g of $MgSO_4.7H_2O$ in sufficient distilled water and make up the volume to one liter).

Calcium chloride solution (Dissolve 27.5 g of anhydrous calcium chloride in sufficient distilled water and make up the volume to one liter).

Ferric chloride solution (Dissolve 0.25 g $FeCl_3$. $6H_2O$ in distilled water and make up the volume to one liter).

Procedure Prepare dilution water by adding 1.0 ml of each of phosphate buffer solution, magnesium sulphate solution, calcium chloride solution and ferric chloride solution to one liter of distilled water. Add 2.0 ml of sewage and aerate. Determine the exact capacity of three 125 ml narrow mouthed flat stoppered bottles. Find out the DO of undiluted sample. Prepare the desired mixture by adding sample in dilution water. Fill up one bottle with the mixture and the other one with dilution water (blank). Incubate at a fixed temperature for a definite time (20 °C-5 days or 35 °C-3 days). Find out DO in both the bottles after incubation.

Table-BOD measurable with various dilutions of samples

Using % mixture (%)	Range of BOD (mg/l)	Direct pipetting into 300 ml bottles (ml)	Range of BOD(mg/l)
0.01	20,000-70,000	0.02	30,000-105,000
0.02	10,000-35,000	0.05	12,000-4,20,000
0.05	4,000-14,000	0.10	6,000-21,000
0.1	2,000-7,000	0.20	3,000-10,500
0.2	1,000-3,500	0.50	1,200-4,200
0.5	400-1,400	1.0	600-2,100
1.0	200-700	2.0	300-1050
2.0	100-350	10.0	60-210
5.0	40-140	20.0	30-105
10.0	20-70	50.0	12-42
50.0	4-14	100.0	6-21
100.0	0-7	300.0	0-7

Calculations

For percent mixtures

$$\text{BOD in mg/l} = \left[\frac{(Dob - Doi) \times 100}{mixture}\right] - (Dob - Dos)$$

For direct pipetting

$$\text{BOD in (mg/l)} = \left[\frac{(Dob - Doi) \times vol.\ of\ bottle}{vol\ of\ sample}\right] - (Dob - Dos)$$

where

Dob - DO of blank (dilution water) after incubation

Doi - DO of mixture after incubation

Dos - DO of undiluted sample

2.14 Chemical Oxygen Demand (COD)

The chemical oxygen demand (COD) of water represents the amount of oxygen required to oxidize all of the organic matter, both biodegradable and non-biodegradable by a strong chemical oxidant. In practice, the organic matter in a sample is oxidized with a known excess of $KMnO_4$. The unused $KMnO_4$ is determined by back titration using potassium iodide and sodium thiosulphate. The amount of $KMnO_4$ consumed in the oxidation of organic matter is then calculated. The amount of $KMnO_4$ consumed may then be easily converted to oxygen equivalent, since 1 mg of potassium permanganate is equal in oxidizing power to 1 meq. or 8 mg of oxygen.

If non-biodegradable organic matter is present, that being the cause with domestic water supplies, then COD>BOD. Typical examples of non-biodegradable organic substances are cellulose, coal dust, lignin, tannin, saw dust, etc. In practice the ratio of COD/BOD varies widely and it cannot be predicted for a given water sample. COD determination has an advantage over BOD test that the results can be obtained in less than 5 hrs. while BOD determination requires 5 days.

Reagents

Potassium permanganate solution	(0.1 N)
Sulphuric acid	(25% V/V)
Potassium iodide	(10%)

Procedure

Take 50 ml of the water sample in a 250 ml conical flask and 50 ml of distilled water in another flask to run the blank. Add 5 ml of $KMnO_4$ (0.1N) solution to both flasks and heat them on water bath at boiling point for one hour and then cool to room temperature. Add 5 ml of potassium iodide (10%) solution in both the flasks followed by 10 ml of H_2SO_4 (25%v/v). Titrate both, sample and the blank with 0.1 N sodium thiosulphate solution using starch as indicator. The blue colour produced by starch would disappear sharply at the end point. Note the reading for samples (S) and blank (B). Repeat the observation 3 times for concordant results.

Calculation

$$\text{COD (mg/l)} = \frac{(B-S) \times N}{Sample\ volume\ in\ ml}$$

Where,

B = the volume of titrant used in blank

S = the volume of titrant used for sample

N = strength of the titrant

2.15 Nitrite-Nitrogen

Principle

Nitrite occurs in water as an intermediate compound in the microbial reduction of nitrate or in the oxidation of ammonia. In addition, nitrite may be excreted by phytoplankton especially during periods of luxury feeding. Nitrite-N is determined through formation of a reddish purple azo-dye produced at pH 2.0 to 2.5 by coupling diazotized sulfanilic acid with N-(I-naphthy)-ethylene diamine dihydrochloride (NED). The applicable range of the method for spectrophotometric measurements is 10 to 1000 mg NO_2-N L^{-1} with 1 cm path length at 543 nm (Greenberg et al., 1985).

Reagents

Sulfanilamide solution (Dissolve 5.0 g of sulfanilamide in a solution of 50.0 ml of concentrated hydrochloric acid and about 300 ml of distilled water. Dilute to 500 ml).

N-(1-naphthyl)-ethylenediamine dihydrochloride solution (NED): (Dissolve 0.5g of the reagent in 500 ml of distilled water. Store in the dark and discard when the solution turns brown)

Standard nitrite stock solution: (Dry pure sodium nitrite at 110 C in desiccator for several hours and cool. Dissolve 49.3 mg of the dried material in deionized water and dilute to 1.0 L to make a 10 mg NO_2-N L^{-1} solution. Add a few drops of chloroform to retard growth of bacteria).

Low concentration working standard: Dilute the standard nitrite solution by adding 10 ml to a liter of deionized water. Each milliliter contains $1x10^{-3}$ mg NO_2-N L^{-1}. This solution should be made fresh each time.

Procedure

Add 2.0 ml each of sulfanilamide solution and N-(1-naphthyl) ethylenediamine dihydrochloride solution to 40.0 ml of sample in a 50 ml volumetric flask. Fill to mark with deionized water. Prepare standards by pipetting into 50 ml volumetric flasks 1.0, 2.0 and 5.0 ml of low concentration working standard as in step 1. Use deionized water to prepare a reagent blank. After 10 minutes but before 2 h, read absorbance against the blank at 543 nm in a cell 1 or 2 cm thick. If absorbance is above 1.0, a smaller sample should be used. Plot mg NO_2-N versus absorbance for the standard solution. Read mg NO_2-N of unknown from the curve and divide this value by the volume in litres to obtain the concentration in mg NO_2-N L^{-1}.

2.16 Nitrate-Nitrogen

Principle

Nitrate is the final oxidation product of nitrogen compounds in water and is considered to be the only thermodynamically stable oxidation level of nitrogen in the presence of oxygen in water. River and pond water, unless contaminated with sewage effluents, city or industrial waste, generally contain quite low amount of nitrate. Some ground water may have appreciable nitrate concentration. Nitrate is reduced to nitrite and the nitrite thus produced is determined by diazotinzing with sulphanilamide and coupling with N-(1-naphthy)-ethylene diamine to form a highly colured azodye that is measured spectrophotometrically.

Reagents

Buffer solution (Mix 20.0 ml of sodium hydroxide solution with 20.0 ml of phenol solution)

Cooper sulphate solution (Dissolve 100 mg of cupric sulphate pentahydrate in deionized water and make up the volume to one liter)

Hydrazine sulphate solution (Dissolve 3.625 g of hydrazine sulphate in deionized water and make up the volume to 500 ml)

N-(1-naphthyl)-ethylene diamine dihydrochloride solution (same as in nitrite test)

Phenol solution (Dissolve 9.072 g of phenol in 200 ml of deionized water)

Reducer solution (Mix 20.0 ml of cupric sulfate solution with 20.0 ml of hydrazine sulfate solution)

Sodium hydroxide solution (Dissolve 28.35 g of sodium hydroxide in deionized water. Cool and dilute to 2 L)

Sulphanilamide solution (Same as in nitrite test)

Procedure

Add 0.5 ml of sample to a 50 ml volumetric flask. Add 20.0 ml of deionized water, 2.0 ml of buffer solution, and 1.0 ml of reducer solution. Let it stand at room temperature for 20 hr. Shield from light. Add 2.0 ml each of N-(1-naphthyl)-ethylene diamine dichloride solution and sulfanilamide solution and fill to mark with deionized water. After 10 minutes but before 2 hours read absorbance against reagent blank at 543 nm in a cell 1 or 2 cm thick. Read milligrams NO_3-N from a standard curve plotted for the nitrite test and divide by the sample volume in litres to obtain mg NO_3^-N L^{-1}.

2.17 Total Ammoniacal Nitrogen (TAN)

Principle

An intensely blue compound indophenol is formed by the reaction of ammonia, hypochlorite and phenol. The procedure employing the Berthelot reaction for ammonia analysis is called indophenol method or phenate method.

2 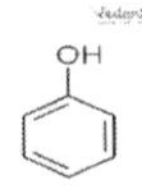 $+NH_3 + 3ClO^- =$ 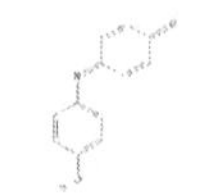 $+ 2H_2O + OH^- + 3Cl^-$

(phenol) indophenol (blue)

Indophenol gives the solution a blue colour, the intensity of which is proportional to the concentration of ammonia present. Both ammonia and ammonium are measured (total ammonia nitrogen), because in a strongly alkaline solution all of the ammonium is converted to ammonia.

Reagents

Phenol-alcohol solution (Dissolve 20.0 g of reagent grade phenol in 200 ml of 95% (vol vol^{-1}) ethyl alcohol).

Sodium nitroferricyanide/Nitroprusside (0.5%). (Dissolve 1.0 g of sodium nitroferricyanide (nitroprusside) in 200 ml of deionized water and store in an amber coloured bottle and refrigerate).

Alkaline solution (Dissolve 100 g of trisodium citrate and 5.0 g of sodium hydroxide in 500 ml of deionized water)

Sodium hypochlorite solution: Use a commercial hypochlorite solution that is at least of 1.5 N).

Oxidizing solution: Mix 100 ml of sodium citrate solution and 25 ml of hypochlorite solution.

Standard ammonia solution (Dissolve 0.472 g of $(NH_4)_2SO_4$ in one liter of deionized water to give a 100-mg TAN L^{-1} solution (100 ppm). To 990 ml of deionized water, add 10 ml of the 100 mg solution to give a 1.0 mg TAN L^{-1} solution. Use 1.0-L volumetric flask. To 45 ml of deionized water, add 5.0 ml of the 1.0 mg solution to give a 0.1 mg TAN L^{-1} solution).

Use $N_1V_1 = N_2V2$ (N=concentration, V=volume; NV=constant)

Procedure

Take 40,0 ml of sample water in a 50-ml volumetric flask. Add 2.0 ml of phenol-alcohol solution in the sample from a pipette. Swirl the flask and add in 2.0 ml of sodium nitroferricyanide solution and 5.0 ml of oxidizing solution. Swirl the flask after each addition. Fill to the mark with deionized water and mix well. Prepare a blank with 40.0 ml of deionized water and follow as above. Pipette in 40.0 ml of standard ammonia solution containing 0.1 mg TAN L^{-1}. Repeat as above. Cover the flask to prevent contamination by atmospheric ammonia. Let it stand for 1 hour at room temperature. Fill a cell of thickness greater than 1.0 cm with blank and set the absorbance at 0.00 for 640 nm. The turbidity blank contains 40 ml of sample diluted to 50.0 ml with deionized water. Use this blank, if necessary. Read the absorbance of the standard, unknown and turbidity blank.

Calculation

$$C_u = \frac{C_s V_s (A_u - A_t)}{V_u A_s}$$

Where,

C_u = the concentration of the unknown in mg TAN L^{-1}.

C_s = the concentration of the standard in mg TAN L^{-1}.

V_u = volume of unknown pipetted in ml.,

A_u = absorbance of the unknown,

A_s = absorbance of standard and

A_t = absorbance of the turbidity blank.

If the absorbance of the unknown is too high, a new sample solution and turbidity blank can be prepared using a smaller volume.

2.18 Available Phosphate

Principle

Ammonium molybdate and potassium antimony tartarate react in acidic medium with orthophosphates to form heteropoly acid which is reduced to intensely coloured molybdenum blue by ascorbic acid.

Reagents

Sulfuric acid solution 5.0 N (Add 70.0 ml of concentrated H_2SO_4 (96%, specific gravity 1.84) in about 500 ml of triple-distilled water and dilute to 1.0 liter).

Ammonium molybdate solution (Dissolve 40.0 g of reagent grade ammonium molybdate $(NH_4)_6 Mo_7O_{24} \cdot 4H_2O$) in about 900 ml of triple-distilled water and dilute to 1.0 liter. Store in a polypropylene bottle in the dark, or in amber coloured, phosphate free, borosilicate glass bottle).

Potassium antimonyl tartarate solution (Dissolve 1.4 g of reagent grade antimony potassium tartarate in triple-distilled water and dilute to 500 ml. Store in a phosphate-free glass bottle).

L-Ascorbic acid solution (Dissolve 1.3 g of reagent grade material in 75.0 ml of triple-distilled water).

Mixed reagent solution (Add 125 ml of sulphuric acid (5.0 N) to a 250 ml flask and add 12.5 ml potassium antimony tartarate solution. Mix thoroughly. Add 37.5 ml ammonium molybdate solution and mix thoroughly. Add 75.0 ml L-Ascorbic acid solution and mix again.

Standard phosphate solution (100 mg PO_4-P L^{-1}) (Dissolve 0.556 g of reagent grade anhydrous (oven dried at 105 °C for 3 hr) potassium dihydrogen phosphate (KH_2PO_4) in distilled water and dilute to 100 ml. Mix thoroughly. Dilute 10 ml of this solution to 1.0 liter with distilled water and add 1.0 ml of chloroform to retard the growth of bacteria. Store in an amber coloured phosphate-free borosilicate glass bottle.

Procedure

Warm the sample to room temperature, then filter through a 0.45µm membrane filter. Pipette in volume of sample likely to contain 0.1 mg PO_4-P into a 50 ml volumetric flask, add 8.0 ml of mixed reagent, dilute to 50 ml with distilled water and mix. Prepare a blank by adding a similar volume of mixed reagent to a 50 ml volumetric flask, dilute to 50 ml with distilled water and mix. Pipette 1.0 ml of standard (0.1 mg PO_4-P) into a 50 ml volumetric flask and add 8.0 ml of mixed reagent, dilute to 50 ml with distilled water and mix. Using the blank and a 2 or 5 cm cell, set the absorbance to 0 at 690 nm. After 10 min but before 3 hours read and record the absorbance of the standard at the above settings. If the absorbance is greater than 1.00, prepare a new solution using a smaller volume. If the absorbance is below 0.2, use a large sample volume. Also prepare a new blank, but not a new standard.

Calculation

$$C_u = \frac{C_s V_s A_u}{V_u A_s}$$

Where,

Cu	=	the concentration of the unknown in mg PO^4-P L^{-1}
C_s	=	the concentration of the standard in mg P_4-P L^{-1}
A_u	=	absorbance of the unknown
A_s	=	Absorbance of the standard
V_u	=	volume of the unknown pipetted in liters and
V_s	=	volume of the standard pipetted in liters.

2.19 Determination of Available Potassium, Calcium and Sodium in water

Determination of Potassium in water

Principle for Potassium Determination

The most common method for K determination is through flame photometry (Pickett and Koirtyohann 1969). The determination is based on measurement of the spectral line intensities of potassium atoms, excited when passing through a flame. Atoms of some specific element (K) take energy from flame and get excited to the higher orbit. Such atoms release energy of a wavelength and give spectral lines which are specific for that element and are proportional to the concentration of atoms of that element.

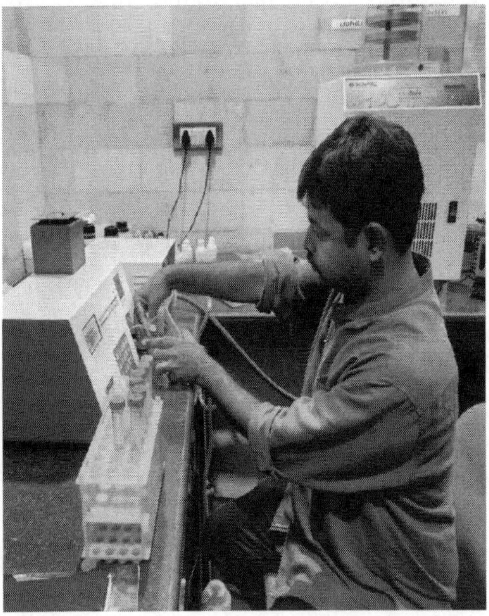

Fig: Use of flame photometer

Reagents

Standard Stock Solution

To prepare a stock solution, 1.9069 g of analytical grade KCl is dissolved in deionized water and volume made up to 1 litre. This solution contains 1000 mg/litre (ppm) K. Prepare 100 mg/litre (ppm) K solution by diluting the 1000 mg/litre (ppm) K solution 10 times (10 ml in 100 mL final volume). Final standard solutions of 0, 5 and 10 mg/litre (ppm) are prepared from 100 mg/litre (ppm) K.

Determination of K in liquid sample

- ☆ For determination of K in liquid sample, first set the instrument with standard solution and prepare standard curve.
- ☆ Standard Curve: A series of standard solution (0, 5, 10 mg/kg) is prepared. The full scale (or 100) of flame photometer is set with solution of highest concentration (i.e., 10 mg/kg). Readings of other standard solution are then taken and curve is plotted.
- ☆ The samples are then read in flame photometer at 548 nm wavelengths or using filter for K.

Calculation

K (ppm) in plant sample can be calculated directly from flame photometer reading or with the help of standard curve.

Calculation for the digested organic material / plant sample

Weight of plant sample = 0.5g

Total volume of plant digest = 50 ml

[In case digestion is carried out in 100 ml Kjeldahl flask the volume would be 100 ml and weight of plant sample would be 1.0 g]

Estimation of calcium using flame photometer

Principal of Flame photometer:

The compounds of the alkali and alkaline earth metals (Group II) dissociate into atoms when introduced into the flame. Some of these atoms further get excited to even higher levels. But these atoms are not stable at higher levels. Hence, these atoms emit radiations when returning back to the ground state. These radiations generally lie in the visible region of the spectrum. Each of the alkali and alkaline earth metals has a specific wavelength. The intensity of the emission is directly proportional to the number of atoms returning to the ground state. And the light emitted is in turn proportional to the concentration of the sample.

Element	Emitted wavelength	Flame Color
Sodium	589 nm	Yellow
Potassium	766 nm	Violet
Barium	554 nm	Lime green
Calcium	622 nm	Orange
Lithium	670 nm	Red

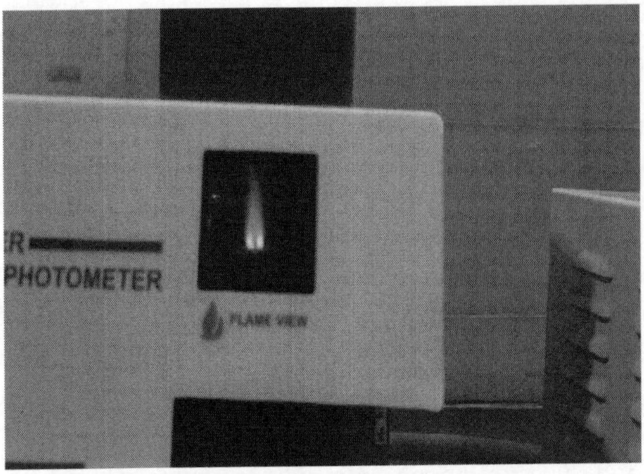

Fig: Orange flame of calcium ions

Chemicals required for this method

- ☆ Calcium carbonate ($CaCO_3$)
- ☆ Hydrochloric acid.(conc. HCl)

Preparation of Solutions

- ☆ Standard Preparation ($CaCO_3$ solution)

Place 1.249 gm AR Calcium Carbonate in approximately 50 ml H_2O. Adding dropwise conc. HCl until calcium carbonate is dissolved (should take about 10 ml).

Procedure

By using $CaCO_3$ we prepared stock solution which has 1249 mg in 50 ml distilled water. Different standard solution can be diluted from 1000 PPM solution of Calcium as follow: 20, 40, 60, 80, 100 PPM. From the standard solution , standard curve is prepared and calcium content in the unknown sample is estimated

Sodium Estimation by Flame Photometer

Flame photometry, more precisely called flame atomic emission spectrometry or "flame photometry" is a traditional and simple method for determining sodium and potassium in sample involves the technique of emission flame photometry. This relies on the principle that an alkali metal salt drawn into a non-luminous flame will ionise, absorb energy from the flame and then emit light of a characteristic wavelength as the excited atoms decay to the unexcited ground state. The intensity of emission is proportional to the concentration of the element in the solution.

As an analytical method, atomic emission is a fast, simple, and sensitive method for the determination of trace metal ions in solution. Atomic emission lines are very narrow (< 0.01 nm). Because of the very narrow and characteristic emission lines from the gas-phase atoms in the flame plasma, the method is appreciably free of interferences from other elements. A photocell detects the emitted light and converts it to a voltage, which can be recorded. Since emit light of different wavelengths (colours), by using appropriate coloured filters the emission due to (and hence their concentrations) can be specifically measured in the same sample. One drawback of flame photometers, however, is that they respond linearly to ion concentrations over a rather narrow concentration range so suitable dilutions usually have to be prepared.

Standard sodium stock solution, 100.0 ppm : Weighing boat. The exact mass was recorded, and corrected the concentrations accordingly. Different concentrations of sodium were prepared and standard curve is plotted. The concentration of the unknown sample is estimated from the standard curve.

2.20 Determination of Trace Elements

The range between the limits of deficiency and toxicity of essential micronutrients is very narrow and hence their estimation is very crucial. Water samples may contain some amount of zinc, iron, manganese, copper, molybdenum and cobalt. These can be determined by atomic absorption spectrophotometer by following the same procedure as described in soil analysis methods using appropriate standards and specific hollow cathode lamps.

Iron Estimation in Water (Spectrophotometer)

Iron occurs in the minerals and leaches to natural water bodies. Elevated iron contents/levels in water can cause stains in plumbing, laundry, cooking utensils, and can impart objectionable taste and colour to food materials. US EPA (United State EnvironmentalProtection Agency) and BIS (Bureau of Indian Standards) approved the iron contents for drinking water i.e. 0.3 ppm.

Materials

- ☆ ***1,10-phenanthroline ($C_{12}H_8N_2$):*** 0.25 per cent solution of the monohydrate in water
- ☆ ***Hydroxylamine hydrochloride ($H_2NOH.HCl$):*** 10 per cent aqueous solution
- ☆ Iron(II) standard solution [$FeSO_4.(NH_4)_2SO_4.6H_2O$]
- ☆ ***Ammonium acetate ((CH_3COONH_4) buffer solution:*** 1 % solution in an acetic acid buffer of pH (mix 65 mL of 0.1 M acetic acid and 35 mL of 0.1 M ammonium acetate solution)

Apparatus Required:

Beaker, Spatula, Measuring cylinder, cuvettes, conical flask, volumetric flask, wash bottle, tissue paper and spectrophotometer (UV-Visible)

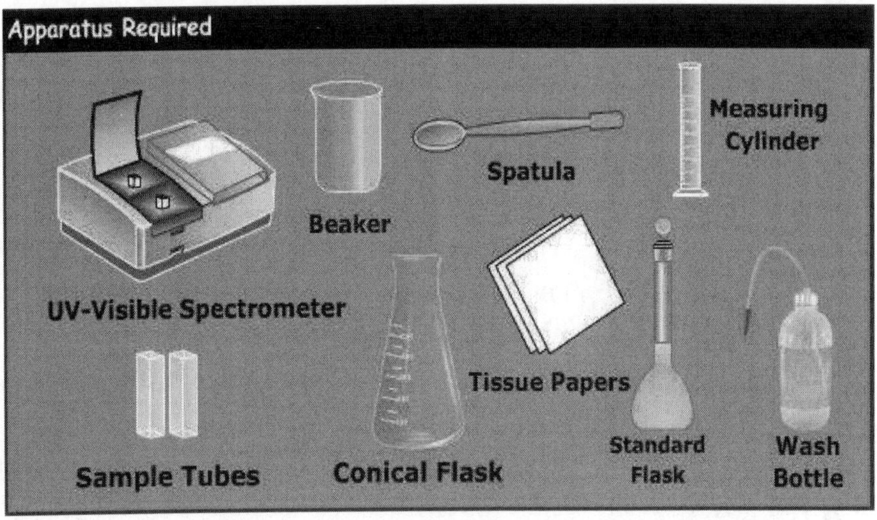

Principles

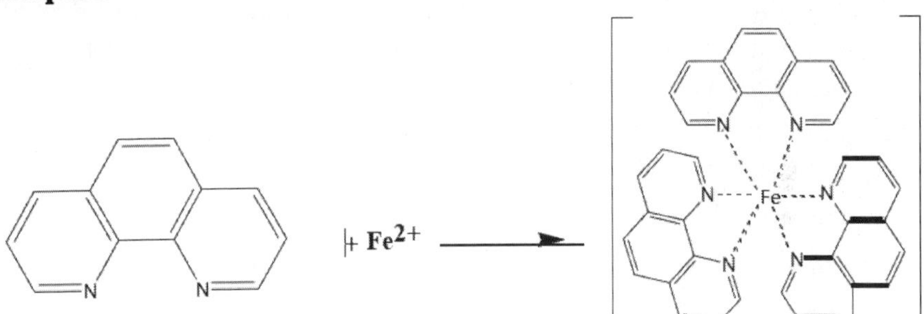

1,10-phenanthroline

Orange-red colour
(Ferrous ortho-phenanthroline-[(C12H8N2)3Fe]2+})

Iron(II) reacts with 1,10-phenanthroline (ortho-phenanthroline) to form an orange-red complex {Ferrous ortho-phenanthroline-$[(C_{12}H_8N_2)_3Fe]^{2+}$} which shows maximum absorbance at wavelength (λ_{max}) 510 nm. The colour intensity is independent of the acidity in the pH range 2-9, and is stable for long periods. First, all the iron (III) contents may be reduced with hydroxylamine solution to iron(II).

Procedure

☆ **Preparation of Stock and Standard Solutions.** Accurately weigh 175 mg of ferrous ammonium sulphate into a clean 500 mL volumetric flask. Dissolve in 100 mL of H_2O and then add 1 mL of concentrated H_2SO_4. Dilute to the mark and mix thoroughly. Calculate the concentration of stock solution, which should be close to 50 ppm Fe.

- ☆ Using a burette, transfer requisite volume of stock iron solution into separate 100 mL volumetric flasks to prepare 10, 20, 30, 40, and 50 ppm solution. Fill each flask about halffull with deionized or distilled water.
- ☆ Add 2mL of conc. HCl solution and 1 mL hydroxylamine solution to each flask, then heat upto boiling to convert all Fe (III) into Fe (II), then cool it.
- ☆ Add 10.0 mL of the ammonium acetate buffer solution to each flask and mix properly.
- ☆ Finally, add 4.0 mL of the 1, 10-phenanthroline solution to each flask. Fill each flask to the 100 mL mark with distilled water.
- ☆ Mix thoroughly and wait for few moments to complete colour development.
- ☆ Similarly blank solution can be prepared: add 10.0 mL of buffer, 2.0 mL of hydroxylamine hydrochloride solution, and 4.0 mL of 1, 10-phenanthroline solution to a 100 mL flask. Dilute to the mark with distilled or deionized water and mix. This is your blank solution.
- ☆ You now have five standard iron solutions and one blank. Record the absorbance of all standard and blank solution by UV-Vis spectrophotometer wavelength max 510nm.
- ☆ Plot the calibration curve between iron concentration v/s absorbance. By using the plot and linear line equation ($y = mx + c$) conversion factor to convert absorbance to concentration can be obtained.
- ☆ Same procedure applied with unknown water sample containing iron and record their absorbance.
- ☆ By correlating with calibration graph, one can calculate iron concentration from unknown water sample.

Results: Given water sample contains iron…ppm.

2.21 Determination of sulfate ion concentration in a water sample

Sulfate is found in appreciable quantity in all natural waters, particularly in arid and semi-arid regions where natural waters generally have high salt content. Sulfate salts are mostly soluble in water and impart hardness. Water with high concentrations has a bitter taste. Sulfate may cause intestinal disorders. These ions can produce hydrogen sulfides as per following equation (1):

SO_4^{2-}+ organic matter - S^{2-}+H_2O+CO_2 (1a) *(in the presence of anaerobic bacteria)*

$S^{2-} + H^+ = HS^-$ (1b)

$HS^- + H^+ = H_2S$ (1c)

The sulfate data is used in determining applicability of different water types for their public and industrial applications. It indirectly indicates extent of problems that can arise due to reduction of sulphates to hydrogen sulphides. In addition, sulphate content of organic matter fed to anaerobic digester is important information as it gives idea of generation of hydrogen sulphides, which need to be removed.

Method:

The turbidimetric method depends on the fact that barium sulfate formed following barium chloride addition to a sample (Equation 2) tends to precipitate in a colloidal form and this tendency is enhanced in the presence of an acidic buffer (consists of magnesium chloride, potassium nitrate, sodium acetate, and acetic acid). These precipitates need to be separated through filtration (using a filter) before sample is analyzed for sulfate concentration. This is a very rapid method and can be used for samples with sulfate concentration greater than 10 mg/L (samples can be diluted and then it can be analyzed).

$Ba^{2+} + SO_4^{2-} = BaSO_4$ (precipitate; poorly soluble) (2)

Reagents:

- ☆ Buffer Solution A: Dissolve 30 g magnesium chloride ($MgCl_2.6H_2O$), 5 g sodium acetate ($CH_3COONa.3H_2O$), 1.0 g potassium nitrate (KNO3), and 20 mL acetic acid (CH_3COOH; 99%) in 500 mL distilled water and make up to 1000 ml.
- ☆ Buffer Solution B (required when the sample SO_4^{2-} <10 mg/L): Dissolve 30 g magnesium chloride, 5 g sodium acetate, 0.111 g sodium sulfate, and 20 mL acetic acid (99%) in 500 ml distilled water and make up to 1000 ml.
- ☆ Dry Barium Chloride ($BaCl_2$) crystals
- ☆ Standard Sulfate Solution: Dissolve 0.1479 g of anhydrous sodium sulfate in distilled water to make the volume 1 L. This solution contains 100 mg sulfate/L (i.e., 1 mL=100 µg $SO4^{2-}$). Prepare standards of various strengths (preferably from 0.0 to 40.0 mg/L at the intervals of 5 mg/L by diluting this stock solution). Above 40 mg/L accuracy decreases and $BaSO_4$ suspensions lose stability.

Apparatus: Whatman No. 1 filter paper; Spectrophotometer; Magnetic stirrer

Procedure:

- ☆ Filter the sample though filter paper (Whatman No. 1) and take 50 ml of filtrate in an Erlenmeyer flask.
- ☆ Add 20 ml buffer solution and mix in stirring apparatus. While stirring,

add 0.15 g of barium chloride to the sample and stir the sample with the help of magnetic stirrer for about an hour.

☆ Measure the absorbance against a distilled water blank (do not add barium chloride to it) at 420 nm using spectrophotometer. Absorbance for the blank sample is taken to correct for sample color and turbidity

Sample Name	Absorbance @420nm	Sample Name	Absorbance @420nm
Distilled water blank		Sample 1	
Standard 1		Sample 2	
Standard 2		Sample 3	
Standard 3		Sample 4	
Standard 4		Sample 5	

2.22 Determination of the chloride ion concentration in a water sample

Chloride in the form of chloride (Cl^-) ion is one of the major inorganic anions in water and wastewater. The chloride concentration is higher in wastewater than in raw water because sodium chloride is a common article of diet and passes unchanged through the digestive system (Average estimate of excretion: 6 g of chlorides/person/day; additional chloride burden due to human consumption on wastewater: 15 mg/l). Along the sea coast, chloride may be present in high concentration because of leakage of salt water into the sewage system. In potable water, the salty taste produced by chloride concentration is variable and depends on the chemical composition of water. Some waters containing 250 mg/l Cl^- may have a detectable salty taste if sodium cation is present. On the other hand, the typical salty taste may be absent in waters containing as much as 1000 mg/l when the predominant cations are calcium and magnesium. In addition, a high chloride contents may harm metallic pipes and structures as well as growing plants. The measured chloride ions can be used to know salinity of different water sources. For brackish water (or sea water or industrial brine solution), it is an important parameter and indicates the extent of desalting of apparatus required. It also interferes with COD determination and thus it requires a correction to be made on the basis of amount present or else a complexing agent, such as $HgSO_4$ can be added. Further, chloride ions are used as tracer ions in column studies to model fate of different contaminants in soil and liquid media.

Method: Mohr Method (Argentometric Method)

The Mohr Method uses silver nitrate for titration (normality: 0.0141) *(method applicability: 0.15 to 10 mg/l chloride ions)*. This corresponds to **1 mL of 0.0141 equals to 1 mg chloride in solution.** The silver nitrate solution is standardized against standard chloride solution, prepared from sodium chloride (NaCl). During the titration, chloride ion is precipitated as white silver chloride (Eq.1):

$Ag^{+}+Cl^{-}$ <=> AgCl (Solubility product constant, Ksp =3 10^{-10}) (1)

The indicator (potassium chromate) is added to visualize the endpoint, demonstrating the presence of excess silver ions. In the presence of excess silver ions, solubility product of silver chromate exceeded and it forms a reddish-brown precipitate (Eq.2). This stage is taken as evidence that all chloride ions have been consumed and only excess silver ions have reacted with chromate ions:

$2Ag^{+}+CrO_4^{2-}$ <=> Ag_2CrO_4 (Ksp=$5x10^{-12}$) (2)

Apparatus: Burette, conical flask, pipette, measuring cylinder

Reagents:

Potassium chromate indicator solution, standard silver nitrate titrant

Procedure:

- ☆ Take 25 ml sample in a conical flask. Measure sample pH
- ☆ Add 1.0ml indicator solution,
- ☆ Titrate with standard silver nitrate solution (0.0141N) to pinkish yellow end point and note down volume of titrant used. Also measure sample pH
- ☆ Calculate chloride ion concentration using Eq.(3):

Chloride Ion Concentration (mg/L) = (A * N* 35.45)*1000/ V_{sample} (3)

Where: A = volume of titrant used, N is normality of silver nitrate (here we used N/71 or 0.0141 N), and V_{sample} is volume of sample used (mL).

Precautions:

1. A uniform sample size must be used, preferably 100 mL, so that ionic concentrations needed to indicate the end point will be constant.
2. The pH must be in the range of 7 to 8 as silver ions are precipitated as AgOH at high pH levels and the chromate ions are converted to Cr_2O_7 2- at low pH values.
3. A definite amount of indicator must be used to provide a certain concentration of chromate ions, otherwise silver chromate may form too soon or not soon enough.
4. Caution should be made to notice indicator color change as it can vary person-to person.

The usual range is 0.2 to 0.4 mL of titrant.

2.23 Sulphide Estimation in Water Sample

Principles

Sulphide often is found in anaerobic waters where it is formed by microbiological reduction of sulphate ions. The un-ionized form, hydrogen sulphide, is highly toxic to fish and other aquatic animals. The un-ionized form exists in a pH and temperature dependent equilibrium with the ionized forms as follows:

$H_2S = H^+ + HS$

$HS^- = H^+ + S^{2-}$

The percentages of un-ionized hydrogen sulphide at different temperatures and pH are provided in tables. The presence of hydrogen sulphide often can be detected by its "rotten-egg" odour. The threshold odor concentration in clean water is between 0.025 and 0.25 µg/litre. This is below the range of analytical detection. The antimony test may be used to verify the presence of sulphide. Add 0.5 ml of a saturated solution of potassium antimony tartrate and 0.5 ml of 6 *N* hydrochloric acid to a 200-ml water sample and shake. The yellow colour of antimony sulphide is a positive test for sulphide. About 0.5 mg/litter of sulphide is necessary for colour formation.

Table : Percentage Un-ionized Hydrogen Sulphide in Aqueous Solution at Different pH Values and Temperatures

pH	Temperature °C								
	16	18	20	22	24	26	28	30	32
5.0	99.3	99.2	99.2	99.1	99.1	99.0	98.9	98.9	98.9
5.5	97.7	97.6	97.4	97.3	97.1	96.9	96.7	96.5	96.3
6.0	93.2	92.8	92.3	92.0	91.4	90.8	90.3	89.7	89.1
6.5	81.2	80.2	79.2	78.1	77.0	75.8	74.6	73.4	72.1
7.0	57.7	56.2	54.6	53.0	51.4	49.7	48.2	46.6	45.0
7.5	30.1	28.9	27.5	26.3	25.0	23.8	22.7	21.6	20.6
8.0	12.0	11.4	10.7	10.1	9.6	9.0	8.5	8.0	7.6
8.5	4.1	3.9	3.7	3.4	3.2	3.0	2.9	2.7	2.5
9.0	1.3	1.3	1.2	1.1	1.0	1.0	0.9	0.9	0.8

The analytical techniques for sulphide measure total sulphide, which includes un-ionized hydrogen sulphide (H_2S), ionized sulphide (HS^- and S^{2-}), and acid-soluble metallic sulphide present in suspended matter. The iodometric method involves the oxidation of sulphide to sulphur by iodine. An amount of iodine solution estimated to be in excess of the sulphide may be added to a sample and the excessive iodine determined by back titration with sodium thiosulfate. This procedure is accurate above about 0.5 mg/litre sulphide, but is subject to interference by reducing substances that react with iodine. However, it is possible to precipitate the sulphide as zinc sulphide by reaction with zinc acetate. The precipitate can then be re-dissolved and the iodometric technique employed with no problem, because interfering substances are removed and the sulphide concentrated in the zinc acetate pre-treatment procedure.

The methylene blue colorimetric method, which is based on the reaction of sulphide, ferric chloride, and dimethyl-*p*-phenylenediamine to produce methylene blue, is sensitive to lower concentration of hydrogen sulphide than the iodometric method. This procedure is much more difficult to conduct than the iodometric technique, and pre-treatment of samples also is desirable to remove possible interference by reducing agents.

Special Apparatus

Needed are narrow-mouthed, glass-stoppered bottles of 500-ml or 1,000-ml capacity; glass filter filtration apparatus, Gelman type A/E glass fibre filters or equivalent; and a micro burette.

Reagents

Zinc Acetate, 2 N : Dissolve 220 g $Zn(C_2H_3O_2)_2 \cdot 2H_2O$ in 870 ml distilled water. This makes about 1 liter of solution.

Sodium Hydroxide 6 *N.* : Dissolve 240 g NaOH and dilute to 1,000 ml with distilled water.

Hydrochloric acid 6 *N.* : Dilute 130 ml of concentrated HCl to 250 ml.

Standard Sodium Thiosulphate solution 0.01N: Prepare a 0.025N solution as described under dissolved oxygen. Dilute the 0.025N solution to obtain a 0.01N solution.

Starch Solution . Prepare as described under oxygen.

Standard Iodine solution 0.025 *N.* : Dissolve 20 to 25 g KI in 200 ml distilled water and add 3.2 g I_2. After I_2 has dissolved, dilute to 1,000 ml. Standardize against 0.0250 *N* sodium thiosulfate solution using a starch indicator as described under oxygen.

Procedure

Pretreatment: Put 1.50 ml of zinc acetate solution into a 1,000-ml glass bottle. Pour sample into bottle by a technique that minimized contact with air. Add 1 ml of sodium hydroxide solution. Insert glass stopper and try to prevent the formation of an air bubble under the stopper. Mix the sample thoroughly. Let the sample stand for 30 minutes for the precipitate to settle. The sample should be stable for at least 6 to 8 hours.

Filter the sample through a glass fiber filter and proceed to the titration step at once.

Titration: Place the filter in a 250-ml Erlenmeyer flask. Add 100 ml of distilled water, use a volumetric pipette to introduce 5 ml of standard iodine solution, and add 2 ml of 6 *N* hydrochloric acid. If the iodine color disappears, add 5 ml more of standard iodine solution. Back titrate with sodium thiosulfate solution (0.0100 *N*). Add a few drops of starch solution near the end point, and the disappearance of the blue color formed by the starch will signal the end point.

Calculations

Use the following equation to compute the total concentration of H_2S, HS^-, and S^2:

Total sulphide in mg/liter as S^2 = $[\{(V_1 \times N_1) - (V_r \times N_r)\} \times 16{,}000\}$/ Sample volume in ml

Where

V_1 = volume of iodine solution in millilitres

N_1 = normality of iodine solution

V_T = volume of sodium thiosulfate solution in millilitres

N_T = normality of sodium thiosulfate solution.

To obtain the concentration of un-ionized hydrogen sulphide, obtain the percentage of un-ionized hydrogen from table for the pH and temperature at the time and place of sampling. The H_2S concertation is computed as follows:

Un-ionized hydrogen sulphide

In mg/liter as S^2 = [total sulphide x (F/ 100)]

Where

F = percentage un-ionized hydrogen sulphide

2. 24 Silica Estimation in Water Sample

Silicon is the most abundant element in igneous rocks and some other types of deposits. Most silica in water is probably derived from the decomposition or metamorphism of silicate minerals rather than from the solution of quartz since quartz is resistant to attack by water. Many waters contain less than 10 mg/1 of silica; those that drain deposits high in silicate minerals, particularly feldspars, often contain up to 60 mg/L concentrations exceeding 100mg/L are not commonly found. The chemistry of silica in solution is not completely understood. It is believed that most silica is present in an un-ionized form, but the silicate ion is undoubtedly present in some waters. Silica is not physiologically significant to humans, livestock, or fish, nor is it of importance in irrigation water.

Molybdate Blue Method

Silica in solution as silicic acid or silicate has the property of reacting with ammonium molybdate in an acid medium to form the yellow-colored silicomolybdate complex. The silicomolybdate complex is then reduced by sodium sulfite to form the molybdate blue color. The silicomolybdate complex may form in water as alpha and beta polymorphs (Strickland, 1952), which have absorbance maxima at different wavelengths. In order to favor the development of the beta form, the pH of the reaction mixture is reduced below 2.5 (Govett, 1961).

The possibility of having larger concentrations of so-called unreactive silica is greater in water containing high concentrations of silica than in water with a low silica content. When significant amounts of unreactive silica are known or suspected to be present, a 1-hr digestion of a 50-ml sample with 5.0 ml of 1.0 M NaOH is suggested as a means of making all the silica available for reaction with the molybdate reagent.

This method may be applied to samples containing less than 100 mg of silica per litre. Samples containing more than 100 mg/1 of silica should be analyzed by standard gravimetric procedures (American Society for Testing and Materials, 1966; Kolthoff and Sandell, 1952).

Interferences

Phosphate gives a similar molybdate complex under certain pH conditions. In the following determination, tartaric acid is added to suppress phosphate interference. There is also evidence that hydrogen sulfide and ferric and ferrous iron interfere with the determination. Hydrogen sulfide may be removed by boiling an acidified sample. The addition of disodium dihydrogen ethylenediamine tetraacetate (Na_2EDTA) eliminates the effect of high concentrations of iron, and also complexes with calcium and prevents precipitation of calcium sulfite.

Reagents

- ☆ Ammonium molybdate solution: Dissolve 52 g $(NH_4)_6Mo_70_{24}.4H_20$ in water, adjust pH to 7-8 with 10M NaOH, and dilute to 1 liter with demineralized water. Filter through 0.45 µm membrane filter if necessary.
- ☆ Hydrochloric acid, 1.0M: Mix 88 ml concentrated HC1 (sp gr 1.19) with demineralized water and dilute to 1 liter.
- ☆ Silica standard solution A, (1.00 ml=0.50 mg SiO_2): Stabilize 3 g of the pentahydrate, $Na_2Si0_3.5H20$, by drying over $CaCl_2$for 2-3 hr. Dissolve 1.765 g in demineralized water and dilute to 1000 ml. The concentration of this solution must be verified by standard gravimetric analysis. Store in a plastic bottle.

- ☆ Silica standard solution B, (1.00 ml =0.005 mg SiO_2) : Dilute 10.0 ml silica standard solution A to 1,000 ml with demineralized water. Store in a plastic bottle.
- ☆ Na_2EDTA solution: Dissolve 10 g Na_2EDTA solution: Dissolve 10 g Na_2EDTA in demineralized water and dilute to 1 liter.
- ☆ Sodium hydroxide solution, 10M: Dissolve 400 g NaOH in demineralized water and dilute to 1 liter.
- ☆ Sodium sulfite solution: Dissolve 170 g Na_2SO_3 in demineralized water and dilute to 1 liter.
- ☆ Tartaric acid solution: Dissolve 100 g $H_2C_4H_40_6$ in demineralized water and dilute to 1 liter.

Procedure

- ☆ Pipette a volume of sample containing less than 0.5 mg Si02 (10.00 ml maximum) into a 50-ml beaker, and adjust the volume to 10.0 ml.
- ☆ Pipette a demineralized-water blank and sufficient standards into 50-ml beakers, and adjust the volume of each to 10.0 ml.
- ☆ To each add, with stirring, 5.0 ml 1.0M HC1, 5.0 ml Na_2EDTA solution, and 5.0 ml ammonium molybdate solution.
- ☆ After 5 min add 5.0 ml tartaric acid solution and mix. Add 10 ml Na_2SO_3 solution and mix.
- ☆ Allow to stand for approximately 30 min. The color is stable for several hours after this time.
- ☆ Determine the absorbance of the test sample and standards against the blank at 700nm
- ☆ Determine mg SiO_2 in samples from a plot of absorbances of standards.
- ☆ SiO_2 in mg/L = (1,000/ml of the sample) *mg SiO_2 in sample

2.25 Phytoplankton Pigment Estimation

Changes in pigment stoichiometry are a key element in the acclimation of plants to fluctuating environmental conditions. The exact evaluation of the eco-physiological meaning of pigment composition requires the separation and quantification of all carotenoids and chlorophylls. A simple method for the extraction, separation and quantification of all major photosynthetic pigments (carotenoids and chlorophylls) is based on an acetone extraction followed by reverse phase chromatography (HPLC) and photodiode array (PDA) detection. It allows a good separation of lutein and its isomer zeaxanthin, and the runtime is relatively short (25 minutes).

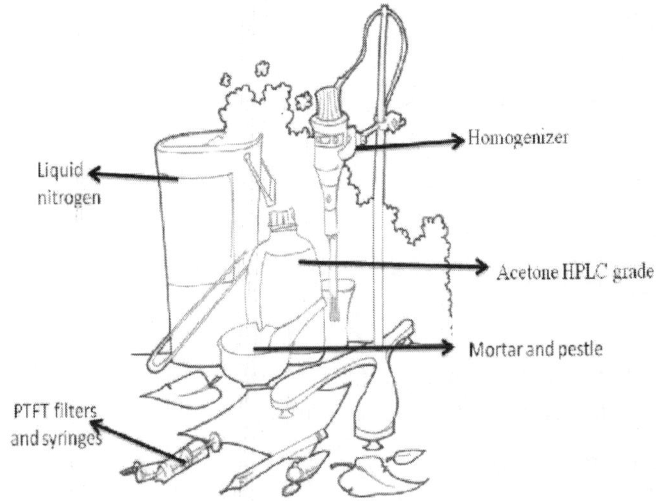

Chlorophylls are not so stable, and high light during the extraction produces phototransformations (standard laboratory illumination is enough to prevent photodegradation of chlorophylls) while the presence of acid traces leads to the formation of pheophytin after the replacement of the Mg^{++} from the chlorophyll ring by H^+. The quantitation of total chlorophylls and carotenoids can be achieved by spectrophotometry, but the separate identification and quantitation of all plant pigments requires the use of an HPLC system. HPLC separates molecules by pumping a mobile phase through a densely packed column. After separation, detection is performed by a photodiode array (PDA) detector that scans the absorbance of the eluent at a range of wavelengths. Each compound (pigment) is detected as a peak of absorbance over the baseline of the chosen wavelength.

Materials/Equipment

- ☆ Mortar and pestle
- ☆ Liquid nitrogen
- ☆ Liquid nitrogen containers
- ☆ Ice containers to maintainthe samples cold
- ☆ Acetone HPLC grade with 0.5 g /L of $CaCO_3$
- ☆ Absolute ethanol
- ☆ Eppendorf tubes
- ☆ Epperdorf centrifuge
- ☆ 1 mL HPLC vials and caps

- ☆ 0.22 µm PTFT filters and syringes (Hydrophobic polytetrafluoroethylene filters, acetone resistants)
- ☆ HPLC system: two pumps, auto sampler, PDA detector (Detection system able to scan a range of wavelenghts simultaneously)
- ☆ Integration software
- ☆ HPLC solvents
- ☆ Pigment standards

Procedure

The Procedure of extraction of pigments from phytoplankton must be adapted to the amount of material, and mode of storage.

Sample collection and storage

Water samples can be collected from the aquatic system at a number of stations. Niskin bottles (5-L) fixed on polyvinyl chloride (PVC)-coated wire for coastal, marine deep water and bucket can be used for inland surface water sampling. For the Phytoplankton pigment analysis, filter 500ml samples immediately after sampling on a GF/F filter (pore size 0.7 µm) on Millipore filtration unit avoiding exposure of the filter paper to direct light and high temperature. The filter paper is was stored in deep fridge at -80^{0} C or in liquid nitrogen until analysed in the laboratory. In general, the universal mode of collection of biological samples implies the use of liquid nitrogen to freeze them and the subsequent storage at -80 °C until analysis. An essential requirement is to maintain the cold chain all the time until the chemical compounds are stabilized with the extraction medium. If samples exposed to light and temperature, even for a few seconds, irreversible chemical modifications may occur, altering the chemical composition. Alternatively, frozen samples in liquid nitrogen can be lyophilized and stored at room temperature. When liquid nitrogen is not available (as in many remote areas), samples can be desiccated and stored in silica as it is described by Esteban et al (2009). Conservation protocols generate two main types of samples: desiccated or frozen (and hydrated). Both types differ in chemical terms by the presence of water. And this water is very important to facilitate the extraction of some xanthophylls, so when water is not present in samples, a small amount is added to the extraction medium. Thus, for frozen samples, which contain water, extraction is performed with pure acetone (buffered with calcium carbonate), but lyophilized or desiccated samples should be extracted with acetone 95-99%.

Extraction using mortar and pestle

- ☆ Chill the mortar with liquid nitrogen and immerse the samples.
- ☆ Homogenise to powder with mortar and pestle. Note that the mortar always must contain liquid nitrogen until acetone is added.

- ☆ Add 1 ml of acetone and homogenize with the sample, the mixture is collected in a 2 mL eppendorf tube. 1 ml of acetone is used to clean the mortar and pool with the extract, the volume in the eppendorf is adjusted to 2 ml.
- ☆ Centrifuge the tube for 5 minutes at a speed of at least 5000 rpm, and collect the supernatant. The Pellet can be re-extracted if it contains visible chlorophyll and pooled together with the first extraction.

Filter extract through a 0.22 µm PTFE filter. The first drops that pass the filter should be discarded to avoid contamination. Fill the HPLC vial and close with a cap. It can be stored inthe freezer for some days but it is better to immediately inject the sample in the HPLC.

- ☆ Clean the mortar, pestle, syringes, filters and other materials with ethanol and dry them with nitrogen gas.

Concentration of Cells and Extraction via Sonication

- ☆ In all steps, avoid degradation of pigments. Minimize exposure of pigments to light and heat.
- ☆ Set up filtering apparatus with glass fibre filters (Whatman GF/F, nominal pore size = 0.7 µm). Cover filter with a small squirt of saturated $MgCO_3$ solution (1 g in 100 mL distilled water).
- ☆ Filter water of suitable volume (in coastal / inland waters, this can be as little as 60 mL; in oceanic, 250 mL to 1 L) using a pressure less than 13 cm Hg.
- ☆ Fold filter, place into centrifuge tube making sure the middle of the filter is facing down, cover with 10 mL of 90% acetone, and cap tightly.
- ☆ Place centrifuge tubes into a cold ice-water bath with the partially submerged sonicator probe. Turn on sonicator probe power for 7 minutes.
- ☆ Let samples sit in the cold and dark for 10 minutes or more. At this point, the samples can be stored in freezer or refrigerator for several days, if necessary.
- ☆ Shake the samples well, otherwise the extract will remain concentrated around the filter and reading will be erroneous.
- ☆ Place samples in centrifuge for 5 minutes at about 5000 rpm (high speed). Ensure that the centrifuge tubes are the correct type for the rotor head.
- ☆ Keep extracts in the cold and dark until reading.

Immerse the frozen filters in 3 mL of 95% acetone (v/v in deionized water) for extraction using a sonicator probe (5 sec, 25 kHz) under low light and temperature (4°C) followed by storage at low temperature (-20°C) for 4 hours. Pass the extract through a Teflon syringe cartridge (Millipore) having a glass fiber pre acrodisc filter (pore size 0.45 µm, diameter 25 mm) to remove the cellular debris. Collect the clarified extract in a 5 mL amber colour glass vial and place directly into the temperature controlled (5°C) auto-sampler trayfor the (HPLC) analysis.

Chromatography

Several methods have been described for pigments analyses. Pigments can be separated following a slight modification of the procedure of Van Hukelem (2002), which provides quantitative analysis of 20 pigments and qualitative analysis of several others. In Fig an HPLC an HPLC system and its components are shown. Basic chromatography is carried out on the HPLC system equipped with an Agilent 1100 pump together with online degasser, an Agilent diode array detector connected via guard column(4 µm; 3.9 20 mm) to an Eclipse XDB C8 HPLC column (5-µm particle size; 4.6x 150mm) manufactured by Agilent Technologies. The column was maintained at 60°C. The solvent flow rate is 1.2 ml min-1 with working pressures below 1000 psi. The injection volume is 15 µl. Elution at a rate of 1.1 ml/minute was performed using a linear gradient program over 22 minutes with 5/95% and 95/5% of solvents B/A being the initial and final compositions of the eluent, where solvent B was methanol and solvent A was (70:30) methanol and 1 M ammonium acetate (pH 7.2) instead of 28 mM solution as recommended in the protocol. An isocratic hold on 95% B was necessary from 22 - 27 minutes for the elution of the last pigment (α and β carotene) at approximately 27 minutes. After returning to the initial condition (5% solvent B) by 31 minutes, the column was equilibrated for 5 minutes prior to next analysis.

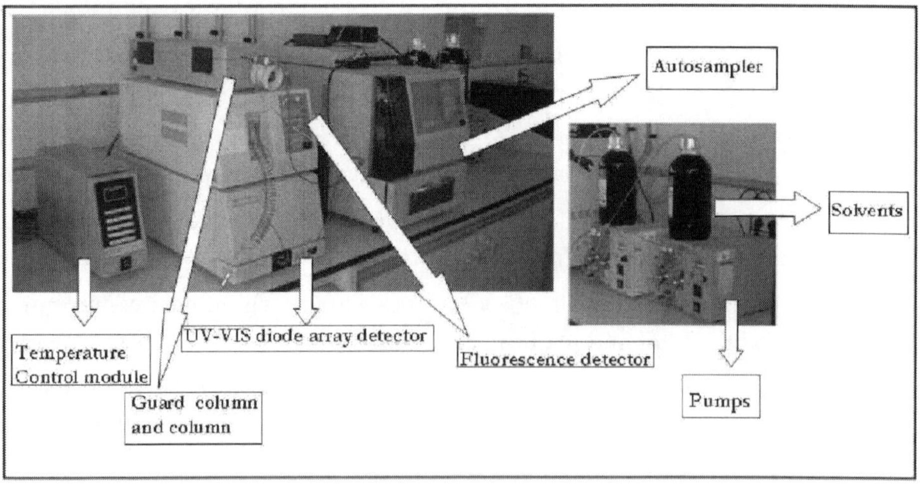

Fig. Typical HPLC system and its components. The system is controlled by a computer.

The eluting pigments were detected at 450 and 665 nm (excitation and emission) by the diode array detector. All chemicals used were of HPLC-grade, procured from E. Merck (Germany). Commercially available standards obtained from DHI Inc (Denmark), were used for the identification and quantification of pigments. Solutions of chlorophyll *a*, chlorophyll *b*, chlorophyll *c2*, chlorophyll *c3*, fucoxanthin, peridinin, neoxanthin, diadinoxanthin, diatoxanthin, alloxanthin, violaxanthin, canthaxanthin, divinyl chlorophyll *a*, -carotene, β- carotene and zeaxanthin were run in order to obtain calibration curves and absorption spectra and to determine detection limits. Identification was based on the retention time and peak shape, i.e. through fingerprint matching with known peak shape from the diode array spectra library created by running the pure standard of individual pigments. The concentrations of the pigments were computed from the peak areas. With the protocol followed, complete resolution and separation of α- and β-carotene could not be achieved; these pigments were, therefore, grouped together and treated as total carotenoids. After pigments separation, a first tentative identification can be done on the basis of the absorption spectra, but the correct identification and quantification requires the use of standards of known concentration. Calibration should be repeated periodically, and it is advisable to add an internal standard (lutein for example) in all sample sets to detect changes in the sensitivity of the detection method. Typically, peaks are detected and integrated at 445 nm for carotenoid and chlorophyll content (Fig.).

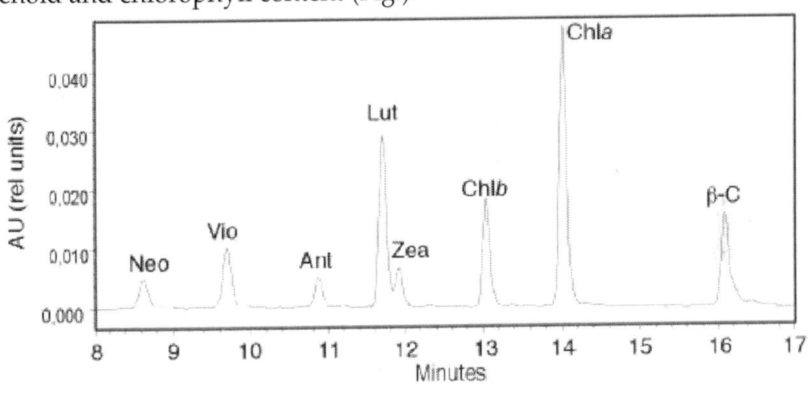

Fig. HPLC chromatograms showing a typical pattern of pigments in an extract from leaves of a typical mesophyte (Taraxacumofficinale) extracted at 445 nm.

The name of each pigment is also shown next to the peak: Neo, neoxanthin; Vio, violaxanthin; Ant, antheraxanthin; Lut, lutein; Zea, zeaxanthin; Chl b, chlorophyll b; Chl a, chlorophyll a, ; β-C, β- carotene.

Table: Retention times and spectral maxima for pigments detected in extracts from plants under the chromatographic system described above

Pigments	Retention times (min)	l max(nm)
Neoxanthin	8.6	437, 466
Violaxanthin	9.7	441, 471
Lutein epoxide	10.6	441, 471
Antheraxanthin	10.9	446, 476
Lutein	11.7	446, 476
Zeaxanthin	11.9	451, 481
Chlorophyll b	13.1	469, 656
Chlorophylla	14.1	435, 666
a-Carotene	16.0	446, 471
b-Carotene	16.2	451, 480

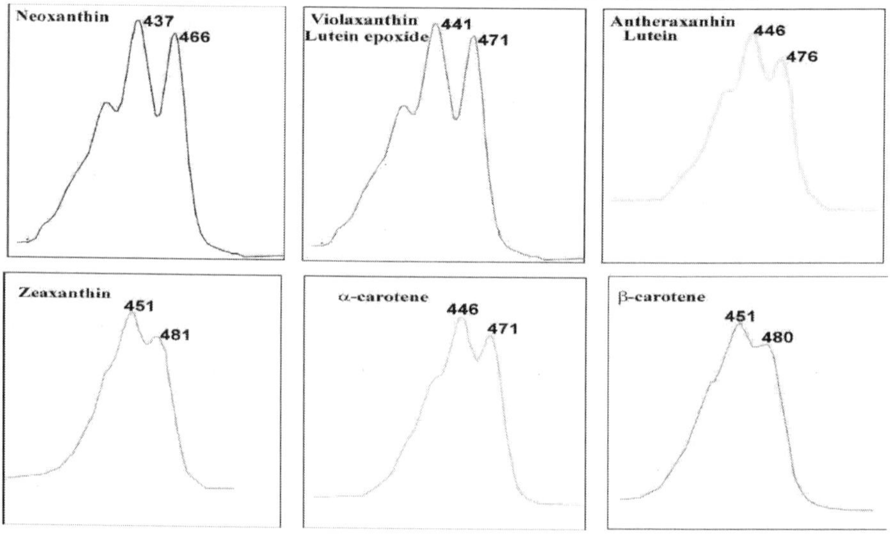

Fig. The spectrum of each carotenoid is shown

Time estimation

The estimated preparation time for the extraction of 24 samples (step 2) is of ~3 hours and ~12 h for the chromatography (step 3)

Usual problems

☆ Samples degraded (exposed to light and high temperature) before the extraction. No solution, sample again.

- ☆ Samples are too tough. Add a hint of sand to the mortar.
- ☆ There is a high amount of pheophytin (a peak that elutes after carotene, its spectrum has two maxima at 410 and 660 nm). This means that there are acid traces in the acetoneor pigment degradation during storage.

Health, safety & hazardous waste disposal considerations

- ☆ Follow the usual safety rules for handling of liquid nitrogen.
- ☆ The organic solvents used in the method have the following risks and hazards:
- ☆ -acetone, acetonitrile, ethyl acetate: highly flammable, harmful methanol: highly flammable, toxic
- ☆ Wear safety glasses and suitable gloves during pigment extraction and solvent preparation.
- ☆ Extraction should be done under a fume hood.
- ☆ Eluent and waste of organic solvents should be collected in a container and disposedof according to the local waste regulations.
- ☆ Store the solvents and waste in fire safe cabinets or in a storage room for such solvents.

2.26 Chlorophyll estimation by spectrophotometer

The concentration of chlorophyll, principally chlorophyll *a*, and after acidification phaeophytin, is measured in 85% acetone extracts using a sensitive fluorometer. The method is very sensitive in the 10 fluorescence units are equal to 0.001 O.D in 10cm cuvette. The maximum variation is 15%. The method is highly suitable for routine analysis at sea or in the laboratory.

Chlorophyll concentration indicates the trophic level of the lake. It is also a measure of the phytoplankton production potential. Though all autotrophic algae contain chlorophyll, the chlorophyll concentration need not be directly correlated with biomass; their relationship is affected, for instance, by illumination, nutrient concentrations and the species composition of the phytoplankton assemblage.

Algal pigments are extracted into organic solvent, in which the concentrations are measured either spectrophotometrically or fluorometrically. Fluorometry may also be used to measure chlorophyll concentrations in vivo in untreated water samples or even in situ. Quantitatively, the most important algal pigment is usually chlorophyll a, which is determined monochromatically. Other chlorophylls (especially bacteriochlorophylls) and their degradation products may absorb partially on the same wavelengths as chlorophyll a. Consequently, the observed concentration may include other pigments in addition to chlorophyll a. Therefore, the spectrophotometric methods always give higher chlorophyll values than those obtained with methods that measure only chlorophyll a (e.g. high performance liquid chromatography HPLC; Meyns et al., 1994).

Ethanol, methanol and acetone have been commonly used in chlorophyll extraction.

APPARATUS AND REAGENTS

i. Spectrophotometer

ii. vaccum filteration device ....

iii. Extraction vessels, e.g. test ....

iv. Refrigerators

v. Solvent

Spectrophotometer

Vacuum filtration device (adjustable vacuum pump, filter holder with clamp, vacuum bottle) Glass-fibre filters free of organic binder (Whatman GF/F 0.7 µm, diameter 47 mm) Glass-fibre filters for filtration of extracts (Whatman GF/F or GF/C, diameter 25mm extraction vessels, e.g.)

Extraction vessels, e.g. test tubes with tight screw caps, capacity of 30-50 ml

Refrigerator

Solvent: Ethanol (C_2H_5OH) 94 % (in terms of volume) / Acetone 95%, N,N-dimethylformamide (DMF/ DMSO) sonication in N,N-dimethylformamide (DMF/ DMSO) should be regarded as a reference extraction method, since it gave the best pigment recovery

Sampling

Vertical chlorophyll profiles should always be taken in connection with primary production measurements. The sampler should preferably be covered with light-impermeable material to protect phytoplankton from direct sunlight. One litre of water (depending on the concentration of cell) for sea sample or 100ml sample from inland system are taken to the laboratory in large plastic bottles, shaded from direct sunlight.

Laboratory Procedure

Chlorophyll samples must be filtered on the day of sampling. Put a 47-mm Whatman GF/F glass-fibre filter on the filter holder, and place the filter system on the mouth of the vacuum bottle. Filter three litres of water through the filter using a suction vacuum of <30 K Pa (<0.3 bar). Low pressure is necessary to avoid pigment losses. If a 1-litre vacuum bottle is used, it must be emptied twice during filtering.

After the filtration is completed (water has disappeared from the surface of the filter), continue filtering for an additional 15—30 seconds to reduce the residual water content of the filter. Remove the filter from the holder, fold (roll) the filter gently with forceps (if this does not help, the filter can also be cut into pieces), and transfer it into an extraction tube. When folding the filter, avoid pressing the upper surfaces tightly together, as this may impair the effectiveness of extraction.

For the extraction of chlorophyll from the algae, the filters are immersed in ethanol. Add 20 ml of 94% ethanol into the extraction tube. Mark each tube to show the sampling location or depth. The extraction tube is closed tightly with its screw cap (instead of cap, parafilm may also be used for air—tight closure of the tubes). The extraction tubes are left overnight in the dark and cold (refrigerator, about +4°C). Next day the tube contents are mixed well, (or can be macerated with sonicator) the filter is removed, and the extract is cleared by filtering it through a Whatman GF/F (or GF/C) 25- mm glass-fiber filter into a clean extraction tube (a filtering syringe may be used). The absorbance of the extract is measured spectrometrically at 665 and 750 nm. The blanks are measured with 94% ethanol. Filtered water and extractant volumes should be selected so as to obtain 665 nm absorbance readings in the range of 0.005 - 0.8.

Do not remove the extract from the cuvette. Add 0.1 ml of 0.3M HCl per 10 ml of extract and mix well. Measure the absorbance again at 665 and 750 nm.

Note

- ☆ If it is impossible to perform chlorophyll extraction immediately after filtration, the filters can be dried in darkness and stored frozen (at least —20°C) maximally for 1 month.
- ☆ If spectrometric determinations cannot be continued immediately after extraction, the cleared extracts can be kept overnight at +4 ^{0} C protected from light (extracts containing the filter must not be stored).

Calculation and Expression of the Result

The chlorophyll concentration of the sample is calculated both with and without acidification. **Chlorophyll estimates without the acidification step are obtained from the equation**

$Cc = ^{Vs}\ 12.0'\ _{Ve} A\ x' \text{-- --- -}$

where

$A = A_{665} — A_{750}$

Cc = chlorophyll concentration of the sample, mg m^{-3} (without acidification)

Ve = ethanol volume used for the extraction,mL

Vs= sample volume filtered in, L

Chlorophyll Estimate Using the Acidification Step is Obtained From the Equation

$$Ca = {}^{Vs\prime} 29.1\prime {}_{d \times Ve} (A- Aa) \prime\text{--- --- - -}$$

where

$A = A_{665} — A_{750}$

$Aa = A_{a\text{-}665} — A_{a\text{-}750}$,

A_{665} = sample absorbance at 665 nm (absorption maximum of chlorophyll a) before acidification

$A_{a\text{-}665}$ = sample absorbance at 665 nm after acidification

A_{750} = absorbance at 750 nm (turbidity) before acidification

$A_{a\text{-}750}$ = absorbance at 750 nm after acidification

Ca = chlorophyll a concentration of the sample, mg m^{-3} (with acidification)

Cc = chlorophyll concentration of the sample, mg m^{-3} (without acidification)

d= cuvette path length, cm

Ve = ethanol volume used for the extraction,mL

Vs= sample volume filtered in, L

12.0 = calculation constant (derived using 83.4 as the absorption coefficient of chlorophyll a in 94% ethanol); if 90% ethanol is used for extraction as in the ISO standard, the value of the absorption coefficient will be 82, which results in a calculation constant of 12.2)

29.1 = calculation constant (derived using an absorbance ratio A/Aa = 1.7 for a solution of pure chlorophyll a which is transformed to phaeophytin by acidification, and the value 83.4 as the absorption coefficient of chlorophyll a in 94% ethanol).

Sources of the Error

Chlorophyll is sensitive to both light and temperature, which should be kept in mind during all phases of the work. Slow disintegration will take place even on frozen filters. The turbidity of the extract will bias the results, so the absorbance reading at 750 nm should not exceed 0.05 with a 50-mm cuvette. The bacteriochlorophylls of autotrophic bacteria (e.g. Chlorobium) absorb at 665 nm, causing high chlorophyll values in some anoxic waters (Takahashi & Ichimura, 1970), even though chlorophyll a may not be present. The possible presence of bacteriochlorophyll in anoxic waters should be checked by plotting the absorption spectrum of the sample and observing the location of the absorption maximum. Chlorophyll results may also be affected by phaeopigments. To detect such interference, additional absorbance measurements of acidified subsamples are sometimes made. The absorbance ratio of non—acidified / acidified samples is 1.7 for pure chlorophyll a samples at 665 nm and lower for samples affected by phaeopigments (Lorenzen, 1967).

2.27 Plankton Cell Counting

Plankton Cell Counting Using a Sedgwick-Rafter Chamber

These counting chambers have been developed for the counting of large cells or where the cultures form long chains or colonies and are suitable where the cell density range is < 10,000 cells / mL. The settling chamber has a volume of 1 mL. To calculate the cell density of a dense algal culture where the cells are relatively small (~ 5-50 μm) and either single cells or short chains, a haemacytometer is more appropriately used.

Method

While the Sedgwick-Rafter chambers are nominally 1 mL, variations in manufacture can create variations in chamber volume. Therefore, it is recommended that prior to use that the chamber volume is calibrated. The volume is calibrated by weight by filling the chamber with deionised water by the Chamber filling method below. The average is calculated from 10 measurements. The chamber measurements of 50 mm long, 20 mm wide and 1 mm deep should also be checked. Any variation from the nominal values that is greater than 5% should be accounted for as a calibration factor in any cell counts.

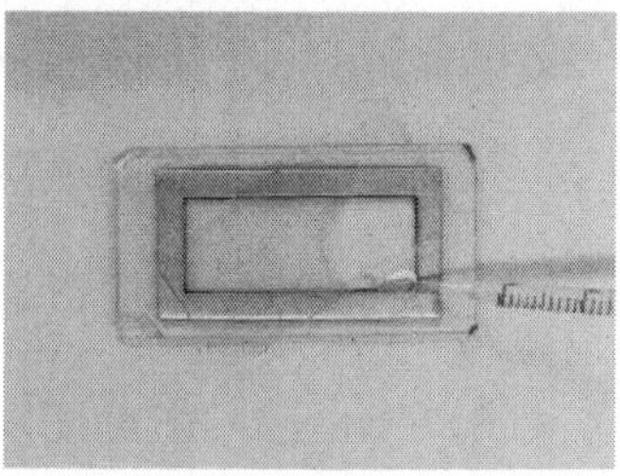

Fig. Filling a Sedwick-Rafter Cell

Chamber filling method: To fill counting chamber, place the thick coverslip diagonally across the cell. Use a pipette to carefully fill the chamber from one corner. Once filled adjust the coverslip to cover the chamber, avoiding bubbles within the chamber.

Algal Sample

Non-motile cells which do not need fixing can be counted as soon as the sample is collected. However, if there will be a delay between sample collection and counting, or if the cells are motile then the sample will need to be preserved.

The most common fixative used for marine microalgae is Lugol solution. The recipe for the acidic form is given here but note in the case of microalgae with calcium carbonate scales, such as the coccolithophorids, the acid will destroy the organisms and a basic solution should be prepared instead.

For cultures add 1 drop of 1-2% Lugol's solution to 1 mL sample, for field samples 10 drops per 200 mL of sample or until the colour of weak tea. Overuse of Lugol's will cause some delicate flagellate species to overstain, lose flagella or blow up entirely. Lugol's is made by dissolving 100 g Potassium Iodide (KI) in 1 L of distilled water, then 50 g crystalline iodine (I_2) is dissolved in this solution and then 100 mL glacial acetic acid is added. Lugol's should be stored in the dark as the iodine is light sensitive and will degrade. It should also be stored with a tight-fitting lid and kept away from the general culture environment.

Mix the sample gently but well and sub-sample the calibration volume of culture with a wide-tipped pipette. Fill the chamber. The time taken to fill the chamber should be short to minimise the settling of cells in the pipette. If flooding or bubbles occur, rinse chamber and coverslip with deionised water, dry with lint free absorbent paper and repeat procedure.

Allow cells to settle for (~15-30 min) or as little time as necessary after focusing through the water column at several points in the chamber to ensure there are no planktonic cells. A benefit of Lugol's is that it increases cell density and shortens settling time. Once the cells have settled, the chamber should also be checked at a low magnification for an even spread of settled cells and to confirm no cells have floated to the top of the chamber. Cultures that are naturally buoyant either from gas-vacuoles (eg. some cyanobacteria) or oil droplets (eg. *Botryococcus*) are difficult to count by this method.

For very dilute samples of large cells or colonies, the entire chamber can be counted to give a cell/colony count. For more dense cultures, the chamber is divided into a grid of 50 squares long x 20 squares width. Cell/colony counts are made for each square for one or two long transects in the chamber. Counting is made to a minimum of 30 counts or until the average and standard deviation of counts per square is stable. Cyanobacteria trichomes can be counted per square as above. Where cyanobacteria cell counts are desired then the number cells per trichome can be counted again for a minimum of 30 trichomes or until the average and standard deviation are stable.

After counting of the chambers, the chamber and coverslip are rinsed with deionised water.

The cell or colony concentration (**C**) to unit per mL for a transect count is calculated using the equation below, followed by adjustments for any dilution or concentration factors.

$$C = (N \times 1000\ mm^3) / (L \times D \times W \times S)$$

Where

N = number of cells/colonies counted

L = length of transect strip (mm)

W = width of transect strip (mm)

D = chamber depth (mm)

S = number of transects counted

Sample Dilution or Concentration

The Sedgwick-Rafter can be used where cell densities are < 10^4 cells / mL. Where culture are very dense and a Haemocytometer is not suitable, such as for some cyanobacteria, it is both inefficient and very difficult to accurately count these cultures without first diluting the sample. Therefore, dilute with a known volume of culture media and then fix.

Alternatively, a Lugol's Dilution Solution (LDS) may be used where 100 mL of 0.22 μm filtered seawater or distilled water (depending on whether the culture is marine or freshwater) is prestained with concentrated Lugols until it is a weak tea colour. Then a known volume of culture can be added to a known volume of the LDS. Using LDS also means that all cells are exposed to the optimum concentration of Lugol's whereas adding concentrated Lugol's could destroy some cell types when it mixes into the sample.

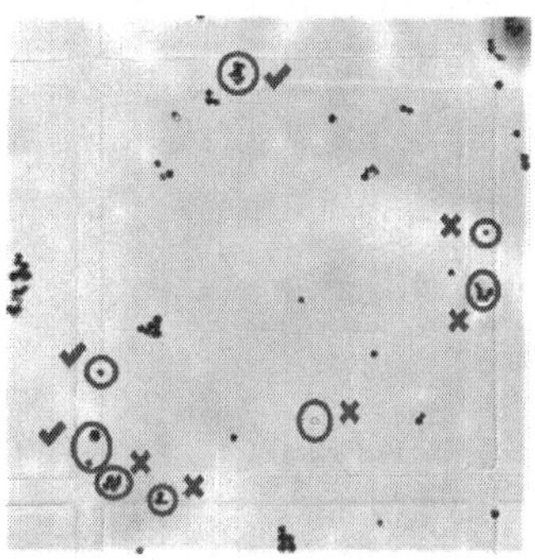

Fig. Sedgwick-Rafter Cell grid indicating edge and empty algal cells to be included/ excluded in counts

If the culture is dilute, concentrate by centrifuging or settling in a flat bottom measuring cylinder (allow 1 hour of sinking for each 10 mm of cylinder height; therefore, overnight is a practical solution). For either method once concentrated, remove and discard up to 90% of the clear supernatant (upper portion of the liquid) without disturbing the settled biomass.

Homogenize the remaining sample and count, bearing in mind the need to integrate the concentration factor = final count x (settled volume / initial volume).

When counting cells bordering on grid lines, convention is to count only those cells touching the top and left-hand side rulings of each square.

2.28 Aquatic Primary Productivity

Principle

The rate at which a given body of water is capable of producing chlorophyll bearing biomass, thus the amount of plant substance produced per unit of time and space is called as rate of gross primary production, and the rate of net primary production represents the rate at which organic matter is stored. The method of estimation of aquatic primary productivity by light and dark bottles was introduced by Gaarder and Graan, in the year 1927. The method consists of taking water samples, containing a natural plankton population in glass bottles and exposing the bottles to light in the euphotic zone. In a parallel experiment a portion of the initial samples is held in a darkened bottle for the same length of time and at same temperature.

The difference between initial O_2 concentration and concentration after a suitable period is a measure of a net evolution of O_2 due to photosynthesis. The amount of the carbon synthesized in the production of glucose during the process of photosynthesis is estimated by knowing the amount of oxygen released as shown in the following chemical equation.

$$\frac{6\ CO_2 + 6\ H_2O\ Chlorophyll \rightarrow C_6H_{12}O_6 + 6O_2}{Sunlight}$$

Materials Required

125 ml capacity D.O. bottles (light and dark), burette, pipettes, conical flask, dropper and plastic rope.

Reagents

Winkler's A (manganous sulphate) solution

Winkler's B (alkaline potassium iodide) solution

Standard sodium thiosulphate (0.025 N or N/40) solution

Starch indicator solution

Concentrated sulphuric acid

Procedure

Fill two light bottles and one dark bottle with water sample. Measure the initial D.O. concentration of one of the light bottles. Suspend the remaining two light and dark bottles with the help of plastic rope at the desired depth, exactly at the depth from where the water samples were drawn. Let the incubation continue, the ideal photoperiod is sun rise to sun set. Estimate the D.O. concentration in light and dark bottles.

Calculation

Gross primary productivity =

i) $\frac{LB - DB}{T} \frac{x}{PQ} 0.375 \times 1000 \; mg\, c / m^3 / hr$

ii) $\frac{LB - DB}{T} \frac{x}{PQ} 0.375 \times 1000 \; mg\, c / m^3 / hr$

Where

IB	:	D.O. Concentration in initial bottle
LB	:	D.O. Concentration in light bottle
DB	:	D.O. Concentration in dark bottle
T	:	Time in hrs. for incubation
PQ	:	Photosynthetic quotient, a numerical constant and its value is 1.2.

2.29 Production of Biochar and its characterization

Production of Biochar

"Biochar is porous carbonaceous residues produced by the pyrolysis of biomass in an oxygen-limited atmosphere" (Shackley et al. 2012). Biochar is dark and light weight, and it contains more carbon. Biochar can be made from different raw materials using a variety of techniques, including pyrolysis, gasification, flash carbonization, and hydrothermal carbonization (HTC).

Pyrolysis Method

Collect the raw biomass of agro waste (sugarcane bagasse, paddy straw) locally from sugarcane processing plant or the farmers field and air-dried till the moisture content reached 12-15% or below. Load the biomass in the electrical heating kiln biomass holder with capacity of 2 kg (for controlled temperature pyrolysis) or the drum kiln (for large production of biochar). The dried biomass was tightly packed in the kiln's biomass holding chamber, the kiln's mouth was closed with a steel lid, and the gas valve was left open to let volatile gases out. While preparing the sugarcane bagasse biochar, the kiln was set to 400°C for 1 hour and 30 minutes, and for the paddy straw biochar, it was 400°C for 45 minutes. After the complete release of volatile gases and cooling of the kiln to room temperature, the cover lid was opened. The biochar yield of bagasse and paddy straw were 33% and 13%, respectively of the initial dried biomass.

Characterization of Biochar

pH and EC

For the pH and Electrical conductivity of biochar, add distilled water to biochar samples (1 g biochar: 20 mL deionized water) and stirring at low speed with a reciprocating shaker for 1.5 hours. The pH of the suspensions was measured using a pH electrode, and the filtrate was used to measure the Electrical conductivity using an EC metre after being filtered using filter paper (Whatman 42mm) (International Biochar Initiative 2015).

Nutrients

For the estimation of the nutrients digest the biochar in a conical flask. 0.5 g of biochar was taken, 10 mL of nitric acid, 2 mL of sulfuric acid were added, and the flask was digested at 150°C until it turned white, and this was utilized for total element nutrient analysis. The leftover digested material was diluted with distilled water and filtered using filter paper. The volumetric flask was filled with 100 ml of filtrate. A total of 10 mL of sample was obtained from the content for total phosphorus reading in a spectrophotometer, and the remainder for potassium measurement in flame photometer and calcium estimation by EDTA titrimetry respectively. Total carbon, nitrogen, hydrogen and sulphur can be estimated by CHNS analyser.

Moisture content, Ash and Volatile Matter

Using the gravimetric technique, the biochar's moisture and ash levels were determined (Novak et al. 2009). To calculate the ash content, the 1 g of biochar was dried in a hot air oven for 24 hours at 105°C before being transferred to a muffle furnace and burnt for 6 hours at 760°C. For the volatile matter content heat the biochar in a muffle furnace at 900°C for 6 minutes and calculate the volatile matter as the percentage of the initial dry weight of biochar. EC values were used to calculate total dissolved solids using the equation TDS = $EC\ (dS\ m^{-1}) \times 640$.

Cation Exchange Capacity (CEC)

Cation exchange capacity (CEC) was quantified according to Song and Guo (2012), with modifications. A quantity of 40 mL of 1 mol L^{-1} ammonium acetate (NH_4OAc) was added to 0.5 g of BC, followed by shaking for 30 minutes; the BC retained in the slow filter was washed with 40 mL of 1mol $L\text{-}^{1}$ NH_4OAc, followed by 90 mL (three portions of 30 mL) of isopropanol. After these washings, 30 mL of 1 mol L^{-1} KCl was added to the samples retained in the filter, which were shaken for 1 hour. This was followed by reading of $NH_4^+$by distillation (Kjeidahl).

2.30 Bio-synthesis of the Nanoparticles

Preparation of Spirulina platensis Micro-Algal Supernatant

Micro-algal powder of Spirulina platensis was prepared by lyphilization of algal cell. The obtained micro-algal powder (MALGP) of Spirulina platensis was initially washed throughout running tap water (TW). Afterwards, they were rinsed three times with DI water to take away any adhered impurities particles. The rewashed MALGP was naturally air-dried for 72 h at ambient temperature (i.e., 25 ±1 °C). The dried MALGP (12 g) was dunked in a 500-mL round-bottomed flask contacting 120 mL of DI water and boiled with continous stirring (~ 150 rpm) using a reciprocal agitator for 1 h at 75 °C to maximize the release of the contained phytochemicals. After allowing the homogenized solution to naturally cool at 25 ±1 °C, the MALGP was screened through a Whatman filter paper (diam. 45 mm) and the supernatant clear solution was collected in polypropylene tubes and kept for the preparation of Spirulina platensis iron oxide nanoparticles (SP-IONPs) or do the centrifugation at 5000rpm for 10 minute to have the supernatant.

Preparation of Spirulina Platensis Iron Oxide Nanoparticles (SP-IONPs)

The best synthesis conditions (maximum absorbance values) were to take 0.6 M of iron (III) of FeCl3.6H2O . It was dispersed in 95 mL of DI water and agitated at ~ 200 rpm for 1 h to ensure its complete dissolution. Then, about 95 mL of MALGP supernatant was gently introduced into the iron (III) suspension (95 mL) with a MALGP supernatant: Fe (III) volume ratio of 1:1. Immediately, the mixture color turned from yellow to intense brown which affirmed the successful synthesis of SP-IONPs. The corresponding solution was left under stirring for another 2 h and the produced homogenous solution was transferred to a hot air oven at 75°C for 24 h. The resulting dark SP-IONPs (solid) was magnetically collected from the solution using magnet, rinsed four times with DI water and dehydrated at 70°C for 6 h before characterization.

Nanoparticle From Seaweed

The green seaweed sample was washed with tap water, followed by double distilled water, and it was ground mechanically. The finely ground algal powder was stored at 20 °C. About 5 g freeze dried sample was boiled with 500 mL double distilled water in an Erlenmeyer flask and stirred continuously for 30 min. The algal extract was cooled to room temperature, filtered using a vaccum filtration unit and stored at 20 °C. Iron nanoparticles were prepared by mixing 25 mL algal extract and 25 mL 0.1 M $FeCl_3$ and the mixture was kept for 2 h at ambient temperature (30 ± 2 °C). The final suspensions were further centrifuged and washed with ethanol (three times) and then dried at 40 °C and nanoparticles were obtained. The aqueous algal extract acts as reducing, stabilizing and capping agent in iron nanoparticles synthesis.

2.31 Lime Requirement

Principles

The general method of (Pillai and Boyd, 1985) is suitable for determining the lime requirement of any soil with exchange acidity. The acid-sulfate soil method is used only with acid-sulfate or potential acid-sulfate soils.

Apparatus

A pH meter is required.

Reagents

Buffer: Prepare a p-nitrophenol buffer of pH 8.0 ± 0.1 by dissolving 20 g of p-nitrophenol, 15 g of H_3BO_3, 74 g of KC1, and 10.5 g of KOH in distilled water and diluting to 1, 000 ml in a volumetric flask.

Procedure

Preparation of Sample: Mud for lime requirement analysis should be collected according to the guidelines given above. Samples may be spread in thin layers on plastic sheets and air dried. Dried mud should be pulverized and the material that passes a 20-mesh screen (0.85-mm openings) used in the test. For most accurate results, the proportion of the entire sample which passes the 20-mesh screen should be estimated and used in establishing the liming rate. However, this step normally is omitted.

Determination of Buffer pH: Weigh 20 g of dry mud (that passing the 0.85-m screen) into a 100-ml beaker, add 20 ml of distilled water, and stir intermittently for 1 hour. Measure the pH of the mud-water mixture with a glass electrode.

Determination of Buffer pH: Add 20 ml of the *p*-nitrophenol buffer to the sample used for the determination of mud-water pH. Stir intermittently for 20 minutes. After setting the pH meter at pH 8.0 with a 1:1 mixture of buffer and distilled water, read the pH of the mud-water-buffer mixture while stirring vigorously.

Calculation

Select the appropriate liming rate from table. If desired, the liming rate can be adjusted according to the proportion of the original mud sample which passed through the 0.85-mm screen.

Comments

If the pH of the mud-water-buffer mixture is below 7.0, repeat the analysis with 10 g of dry mud and double the liming rate from table.

Table: Lime Requirement in Kg/Hectare of Calcium Carbonate (Neutralizing Value of 100) to Increase Total Hardness and Total Alkalinity of Pond Water to 20 mg/liter or above

Mud pH in water	Kg/hectare of calcium carbonate required according to mud pH in buffer solution									
	7.9	7.8	7.7	7.6	7.5	7.4	7.3	7.2	7.1	7.0
5.7	91	182	272	363	454	544	635	726	817	908
5.6	126	252	358	504	630	756	882	1,008	1,134	1,260
5.5	202	404	604	806	1,008	1,210	1,411	1,612	1,814	2,016
5.4	290	580	869	1,160	1,449	1,738	2,029	2,318	2,608	2,898
5.3	340	680	1,021	1,360	1,701	2,042	2,381	2,722	3,062	3,402
5.2	391	782	1,172	1,562	1,953	2,344	2,734	3,124	3,515	3,906
5.1	441	882	1,323	1,765	2,205	2,646	3,087	3,528	3,969	4,410
5.0	504	1,008	1,512	2,016	2,520	3,024	3,528	4,032	4,536	5,040
4.9	656	1,310	1,966	2,620	3,276	3,932	4,586	5,242	5,980	6,552
4.8	672	1,344	2,016	2,688	3,360	4,032	4,704	5,390	6,048	6,720
4.7	703	1,412	2,116	2,822	3,528	4,234	4,940	5,644	6,350	7,056

2.32 Ideal values of various Physico-chemical parameter for Aquaculture in three types of water bodies

S. No.	Parameter	Fresh water	Brackish water	Seawater
1.	Colour (colour units)	Clear water with greenish hue <100 colour units	Clear water with greenish hue < 100 colour units	Clear water with greenish hue <100 colour units
2.	Transparency (cm)	20-35	26-35	26-35
3.	Clay Turbidity (mgl-1)	<30	<30	<30
4.	Solids (mg l -1) a. Total b. Suspended	<500 30-200	<500 25-200	<500 25-200
5.	Temperature (°C) Tropical climate Temperate climate	25.32 10-12	25.32 10-12	25.32 10-12
6.	pH	6.7-9.5	7.0-8.7	7.0-8.5
7.	Hardness (mgl-1)	30-180	>50	>50
8.	Alkalinity (mg l-1)	50-300	>50	>50
9.	Chlorides (mg l-1)	31-50	>500	>500
10.	Salinity (ppt)	<0.5	10-25	>30
11.	D.O (mg l-1)	5-10	5-10	5-10
12.	Total dissolved free carbon dioxide (mg l-1)	<3	<3	<3
13.	Ammonia Nitrogen (NH3-N), mg l-1) Unionized (NH3) Lionized (NH4+)	0-0.1 0-1.0	0-0.1 0-1.0	0-0.1 0-1.0

S. No.	Parameter	Fresh water	Brackish water	Seawater
14.	Nitrite Nitrogen (NO2-N), (mg l-1)	0-0.5	0-0.5	0-0.5
15.	Nitrate Nitrogen (NO3-N), (mg l-1)	0.1-3	0.1-3	0.1-3
16.	Total Nitrogen (mgl-1)	0.5-4.5	0.5-4.5	0.5-4.5
17.	Total phosphorous (mg l-1)	0.05-0.4	0.05-0.5	0.05-0.5
18.	Potassium (mg l-1)	0.5-10	>0.5	>0.5
19.	Calcium (mg l-1)	75-100	>75	>75
20.	Magnesium (mg l-1)	20-200	200-1350	>1350
21.	Sulphate (mg l-1)	20-200	200-885	885
22.	Silica (mg l-1)	4-16	>5	>5
23.	Iron (mg l-1)	0.01-0.3	0.01-0.3	0.01-0.3
24.	Manganese (mg l-1)	0.001-0.002	0.002-0.02	0.002-0.02
25.	Zinc (mg l-1)	0.002-0.01	0.002-0.01	0.002-0.01
26.	Copper (mg l-1)	0.003-0.005	0.003-0.005	0.003-0.005
27.	Cobalt (mg l-1)	<0.003	<0.003	<0.003
28.	Biochemical oxygen demand (mg l-1)	<10	<15	<15
29.	BOD load (kg/ha/day) For sewage fed fisheries	150	100	75
30.	Chemical oxygen demand (mg l-1)	<50	<70	<70
31.	Hydrogen sulphide (mg l-1)	<0.002	<0.003	<0.003
32.	Residual Chlorine (mg l-1)	<0.003	<0.003	<0.003
33.	Primary productivity (mg c/ m3/day)	1000-3000	1000-2500	1000-2500
34.	Plankton (ml/100 liter)	2	1	1
35.	Chlorophyll – a (µg l-1)	20-275	20-250	20-250
36.	Redox potential (volts)	0.40-0.52	0.40-0.52	0.40-0.52

2.33 Terminologies Commonly used in Soil and Water Analysis

Standard solution

A standard solution is a solution of accurately known concentration prepared from a primary standard (a compound which is stable, of high purity, highly soluble in water and of a high molar mass to allow for accurate weighing) that is weighed accurately and made up to a fixed volume.The concentration of the solute can be expressed as normality or molarity. The standard may be a primary or secondary.

Normality

As per the standard definition, normality is described as the number of grams or mole equivalents of solute present in one litre of a solution. When we say equivalent, it is the number of moles of reactive units in a compound.

$$N = \frac{No.\ of\ grams\ of\ solute\ per\ litre}{Gram\ equivalent\ weight}$$

Molarity of the Solution

Molarity of a given solution is defined as the total number of moles of solute per litre of solution. The molality of a solution is dependent on the changes in physical properties of the system such as pressure and temperature as unlike mass, the volume of the system changes with the change in physical conditions of the system. Molarity is represented by M, which is termed as molar. One molar is the molarity of a solution where one gram of solute is dissolved in a litre of solution.

Molarity Formula

The equation for calculating molarity is the ratio of the moles of solute whose molarity is to be calculated and the volume of solvent used to dissolve the given solute.

$$M = \frac{n}{v}$$

where,

M is the molaity of the solution that is to be calculated

n is the number of moles of the solute

V is the volume of solution given in terms of litres

Molality

Molality is also known as molal concentration. It is a measure of solute concentration in a solution. The solution is composed of two components; solute and solvent. There are many different ways to express the concentration of solutions like molarity, molality, normality, formality, volume percentage, weight percentage and part per million. The term needs to calculate the mass of the solvent and moles of solute.

Formality :

Formality is the **number of grams of formula masses of solute dissolved per litre of solution** is known as the formality of a solution. e.g. 1 F, Na_2CO_3 is 106 g of Na_2CO_3 in 1 litre solution. In other words the formula weight dissolved in 1000 ml of the solution is called Formalty. The sign F represents the formality of a solution.

Percentage Strength

The no. of grams of the substance present in 100 ml of the solution is called percentage strength of the solution.

Precision

Precision of a numerical result is the degree of agreement between it and other values obtained under substantially the same condition. It is a measure of the degree to which results "check" Precision describes the reproducibility of values. The closeness of two or more measurements to each other is known as the precision of a substance. If you weigh a given substance five times and get 3.2 kg each time, then your measurement is very precise but not necessarily accurate.

Accuracy

Accuracy is a measure of how close a measurement is to the correct or accepted value of the quantity being measured. Precision is a measure of how close a series of measurements are to one another. Analytical results that are of a high degree of precision are not necessarily of high degree of accuracy. It refers to the reliability of results e.g. when A, B and C determine the weight of a substance, if A records the weight as 18.0 g, B as 18.2 g and C as 19.5 g and the actual weight is 20 g then C is more accurate whereas the determinations of A and B are more precise. Precise determinations may not be accurate.

PRIMARY STANDARDS

Standards prepared from primary chemicals are called as primary standards. A primary standard is a reagent that is extremely pure, stable, has no water of hydration and has a high molecular weight. Some primary standards are sodium carbonate, potassium hydrogen iodate, potassium dichromate, oxalic acid etc. Primary chemicals are substances which are obtained either from nature or by artificial means in comparatively pure form and when a weighed quantity of these chemicals are dissolved in a known volume of a solvent, solutions of definite and stable strength are obtained

Example of primary chemicals

- ☆ **Acids:** Potassium hydrogen phthalate ($C_8H_5O_4K$), Succinic acid ($C_4H_6O_4$); Benzoic acid (C_6H_5COOH) and Sulphamic acid (H_3NSO4)
- ☆ **Bases:** Sodium carbonate (Na_2CO_3), Mercuric oxide (HgO) and Borax ($Na_2B_4O_7$ $10H_2O$)
- ☆ **Oxidising agents:** Potassium dichromate (k_2Cr2O_7) and Potassium iodate (KIO_3)
- ☆ **Reducing agents:** Sodium oxalate, Oxalic acid and Pure iron metal
- ☆ **Neutral salts** – Sodium Chloride (NaCI) and Potassium Chloride (KCI)

Criteria for primary standards

1. Purity > 99.95%
2. High equivalent weight

3. High dissociation constant (acids)
4. Crystalline and solid
5. Should withstand drying temperature
6. Soluble in H_2O and Organic solvents
7. Should have an inertness

A chemical which satisfies majority of these qualifications can be taken as primary standard:

Purity: Permissible limit of impurity is 0.04%. If SO_2 is present in silver nitrate it is not suitable because it will reduce $AgNO_3$ to Ag which will be blackened. When $KMnO_4$ contains MnO_2 it is unsuitable for a primary standard When Na_2CO_3 contains $NaHCO_3$ as impurity it can be eliminated by heating.

Equivalent weight: The gram equivalent of the substance should be as large as possible When the equivalent weight is high the error in calculation of normality will be less.

Dissociation constant for acids: It should be high so that the distribution will be rapid and uniform. However, the dissociation constant of organic acids is very less.

Crystalline and Solid: When the substance is crystalline, solid handling will be easy. It will withstand drying temperature.

Generally the moisture is removed at 105-110°C. So a primary standard should not decompose at this temperature. Na_2CO_3 can withstand up to 260°C and Na oxalate upto 100 °C.

Solubility: Primary standard should be soluble in water (hot water/non polar solvents/mixture of polar and non-polar solvents), e.g. Benzoic acid is insoluble in water but soluble in water + alcohol. When the substance is inert its storability will be long.

INDICATOR

Indicators are chemicals used in volumetric analysis to indicate the end point of reaction. Chemical indicator, is any substance that gives a visible sign, usually by a colour change, of the presence or absence of a threshold concentration of a chemical species, such as an acid or an alkali in a solution. An example is the substance called methyl yellow, which imparts a yellow colour to an alkaline solution. They are classified as self indicators b) Internal indicators and c) External indicators

Self Indicators

Any one of the two solutions involved in the titration will produce a colour change at the end point e.g. Titration involving $KMnO_4$

Internal Indicators

These indicators are directly added to the solution being titrated. These are acted upon only at the end of the reaction, e.g. phenolphthalein, methyl red etc.

External Indicators

These are not directly added. If they are added directly they will be acted upon immediately by the reagents involved. In this case a few drops of the solution from the titration flask are taken out from time to time and tested outside with the indicator solution for completion of the reaction, e.g. Potassium ferricyanide in the estimation of Fe by titration with $K_2Cr_2O_2$

Mixed Indicators

In some titrations involving weak acids vs. weak bases, the end point cannot be easily detected by a single indicator. In such cases one indicator is mixed with another indicator or an indicator is mixed with an organic dye. Either one or both of them may be sensitive to pH. Such mixed indicators enable us to detect the end point more accurately even within a narrow range of pH, e.g. methyl green and phenolphthalein = pH 8.4 – 8.8 (Red-violet at pH < 8.4. grey pale blue at pH 8.4 -8.8 and pink above pH 8.8

Universal Indicators

A universal indicator is a mixture of different types of indicators that exhibits different coloration at different levels. It is used to detect the acidic or basic nature of a substance or a solution. It can be in the form of a paper strip or a solution. Example: methyl red and phenolphthalein. These are not employed in quantitative analysis. These are employed to determine the approximate pH range of a solution. They are available in two ranges *viz,* narrow where the colour change is in the pH range of 0.2 to 0.4 units and wide where the colour changes in 2 pH units say 6-8, 8-10.

pH/Test Papers

These are filter papers impregnated with indicators. They serve a similar purpose as that of universal indicators. They are also in two ranges namely narrow and wide.

Indicators may also be neutralisation indicators (Acid – base), Redox indicators e.g. Diphenylamine, Metal ion indicators e.g. Murexide, Eriochrome black-T, and Adsorption indicators e.g. Potassium chromate in chloride estimation.

Theory of Indicators

Ostwalds (1894) theory: Also referred to as the ionic theory of indicators.

According to this theory, "neutralization indicators are weak organic acids or bases in which undissociated molecules differ in colour from their ions".

eg. Phenolphthalein is a weak acid that does not dissociate in an acid medium and hence has no colour with the acid solution. In an alkali medium, the molecules dissociate and combine with alkali to produce a pint coloured salt.

$HPh \rightarrow H^+ Ph$

$NaOH \rightarrow Na^+ + OH$

$Ph + Na^+ \rightarrow Ph - Na$ (Na phenolphthate).

$Ph - Na \rightarrow Ph\ Na^+$ (pink colour)

Similarly, methyl orange which is yellow in unionised conditions produces red colour when ionised. The undissociated molecules of basic indicators can be denoted by Ind OH and their cation by Ind+. They dissociate as

$Ind\ OH \rightarrow Ind^+ + OH-$

According to this theory, the colour change of indicators is not due to ionisation but due to a change in the structure of indicators. The colour change of indicators is due to some intra-molecular rearrangement which changes the structure of the indicators.

The colour of organic compounds is due to the presence of chromophores in their molecules. These chromophores may be radicals (atomic groups) or groups of double bonds. These include

- ☆ Nitro groups O = N = 0 which can be converted into 0 = N – OH II group
- ☆ Azo group :- N = N- which may be converted into = N – NH- group
- ☆ Quinoid group transformed from benzenoid Carbonyl group C = 0
- ☆ Double bonds close to each other

The colour of organic compounds is also due to the presence of auxochromes. These cannot themselves produce colours but when present along themselves.

Present along with chromophores they augment the action of the latter and deepen the colour produced by them. Eamples of auxochromes – OH, – NH_2, OCH_3, -N $(CH_3)_2$ etc.

According to this chromophore theory the colour change in indicators is the result of Tautomerism which is a reversible interconversion of isomeric forms. Due to this intrramolecular regrouping when the structure changes with the formation or disappearance of groups which influence colour namely chromophores and auxochromes the colour of indicators also changes. e.g. paranitrophenol.

The formation of quinoid nucleus under alkaline condition is the cause for colour change of paranitrophenol under alkaline condition. When it is acidified the equilibrium is shifted in the opposite direction and the colour changes from yellow to colourless. The isomers, involved in tautomerism are called a tautomers.

Choice of Indicator

Two quantitative methods are available for choice of indicators.

- ☆ Calculation of indicator error
- ☆ Plotting of titration curves

Calculation of Indicator Error

The colour of indicator changes over a narrow range of pH and this range depends only on the properties of the given indicator and does not depend upon the reacting acid or base. Due to this the colour changes of an indicator generally takes place not at the equivalence point but with deviation. Such deviation is called as indicator error. It is the error caused by the difference between the titration exponent of a given indicator and the pH at equivalence point. Several methods are available for calculating indicator error.

Plotting of Titration Curves

Strong acid Vs. Strong Base

Titration Curve: 0.1 N HCl Vs 0.1 N NaOH

The salt formed does not hydrolyse. Hence the pH is at equivalence 7.0. In such cases the indicator with a titration exponent (pT) of 7.0 is suitable. The titration exponent (pT) is the pH at which the titration in the presence of a given indicator is ended.

Region of abrupt change is between 4.00 and 10.0. Hence indicators with pT of 4.0-10 can be used i.e. methyl orange – phenolph-thalein.

The indicator used must have titration exponent within the region of abrupt change in the titration curve.

Weak Acid vs Strong Base

Titration: 0.1 N CH_3 COOH vs 0.1 N NaOH

The salt formed is alkaline. The pH at equivalence is > 7.0. The equivalence point.

The Equivalence point lies in the alkaline range. The break of pH in the curve is at pH 7.73-10.0. Phenolphthalein is the best indicator.

Weak Base vs. Strong Acid.

The salt formed is acidic.

The Equivalence point is in the acid range (pH 5.12). The region of abrupt change is 4.0 and 6.2. Hence suitable indicators are methyl red and methyl orange.

Weak Acid vs. Weak Base

There is no pH break. This means that titration cannot be performed accurately with any of the known indicators. Hence in titration by neutralisation method one of the reacting substances must be a strong electrolyte.

BUFFER SOLUTIONS

A buffer solution is that which resists change in pH on the addition of small amounts of acid or alkali or by dilution. If a buffer is introduced into a reaction system, the solution pH remains almost the same in spite of the formation of acid or base in the reaction.

Properties of Buffer

- ☆ It should have a definite pH at a particular concentration
- ☆ The pH should be practically unchanged if diluted.
- ☆ The pH should slightly change by the addition of strong acid or strong base.

Preparation of Buffer Solutions

Commonly prepared by mixing an acid or base (usually weak) with its salt

Eg. $CH_3COOH + CH_3COONa$ – Acetic acid + Sodium acetate

$NH_4OH + NH_4CI$ – Ammonium hydroxide + ammonium chloride

There are different types of buffers. Each maintains a fixed pH at a particular concentration.

Buffer Action

Eg. The action of a strong acid like HCI with a buffer say $HCOONH_4$

$$HCOOH + HCOONH_4 + HCI \rightleftharpoons NH_4CI + HCOOH$$

$$HCOO^{-1} + H^{+} \rightleftharpoons HCOOH$$

The strong acid reacts with formate and produces a weak acid namely formic acid. Practically there is no change in pH.

Similarly when an alkali is added the OH ions of alkali combine with the H ions of the formic acid and the pH remains the same

These buffers function so long, the reaction between buffer components and acid/bases is incomplete. When the reaction is complete further addition of acid/ base changes the pH appreciably and the buffer system is destroyed.

APPENDICES

Appendix I

Conversion factors:

meq	mg/eq.wt
ppm	mg/l or µg/ml
1 ppm	2.25 kg/ha
10,000 ppm	1 per cent
1 mg/100 g	22.5 kg/ha
1 kg/ha	0.1 g /sq m
1 hectare	2.471 acres
1 acre	0.405 hectare
1 kg /ha	1.12 lbs/ acres
Mesh number	16/ mm
Normality	(Specific gravity X Purity X 10 (Liquid))/Equivalent weight
Normality	Wt of salt per liter X Purity (Solid)/ Equivalent weight
Grams per liter	Normality X Eq. Wt
Organic matter	Organic carbon X 1.724
Optical density	2 – log T (T= transmission)
Per cent by Wt	Grams of solute/ 100 g of solution
1 Angstrom (Å)	10-8cm or 10-10 m
10 (Å)	1 nanometer or 1 millimicrometer

Temperature Conversion

°C / 5	(°F-32) / 9
P X 2.29	P_2O_5
P_2O_5	0.44 X P
K X 1.20	K_2O
K_2O X 0.83	K
S X 3	SO_4
N X 1.12	NH_3
Protein (%) Nitrogen (%) X 6.25	
Velocity of light	3 X 10^{-10} cm / sec.
Velocity of sound	332 m / sec.

Appendix II

Percentage composition of manures and fertilizers

Material	N	P_2O_5	K_2O	Others
FYM	0.5-1.5	0.4-0.8	0.5-1.9	-
Compost (urban)	1.0-2.0	1.0	1.5	-
Green manure	0.5-0.7	0.1-0.2	0.8-1.6	-
Bone meal(steamed)	1.0-2.0	25-30	-	-
Anhydrous ammonia	82	-	-	-
Ammonium chloride	24	-	-	-
Ammonium nitrate	33	-	-	-
Ammonium sulphate	20.6	-	-	24 (S)
Ammonium phosphate	11	48	-	-
Diammonium phosphate	21	53	-	-
Calcium cynamide	20	-	-	-
Calcium nitrate	16	-	-	-
Sodium nitrate	16	-	-	-
Urea	46	-	-	-
Super phosphate	-	16	-	12 (S)
Dicalcium phosphate	-	23	-	-
Tricalcium phosphate	-	45	-	-
Rock phosphate	-	11-17	-	-
Basic slag	-	3.5-8	-	-
Muriate of potash	-	-	60	-
Sulphate of potash	-	-	48	18 (S)

Material	**N**	**P_2O_5**	**K_2O**	**Others**
Zinc sulphate	-	-	-	35 (Zn); 18 (S)
Zinc chelate	-	-	-	14 (Zn)
Copper sulphate	-	-	-	25 (Cu); 13 (S)
Ferrous sulphate	-	-	-	19 (Fe); 19 (S)
Iron chelate	-	-	-	5-9 (Fe)
Borax	-	-	-	11 (B)
Sodium tetra borate	-	-	-	14 (B)
Manganese sulphate	-	-	-	26 (Mn)
Manganese chelate	-	-	-	10-12 (Mn)
Calcium sulphate (Gypsum)	-	-	-	18 (S); 33 (Ca)

Appendix III

General Recommended Doses of Micro-nutrient Fertilizer

Micronutrient	Material and doses for application	
Zinc	Zinc sulphate (25-50 kg ha-1) (25 kg ha-1 for light soils and 50 kg for heavy soils)	0.5% zinc sulphate + 0.25%lime
Iron	Ferrous sulphate (25-50 kg ha-1)	1-2% ferrous sulphate + half of lime
Copper	Copper sulphate (10 kg ha-1)	0.1% copper sulphate + 0.05% lime
Manganese	Manganese suphate (10 kg ha-1)	1% manganese sulphate + 0.25% lime
Boron	Borax (10 kg ha-1)	0.2% borax

Appendix IV

Classification of Pond Productivity

Pond productivity	**Organic carbon (%)**	**Available N**	**Available P**
		(mg/100g soil)	
Low	< 0.5	<25	<3
Medium	0.5 – 1.5	25-30	3-6
High	>1.5	>50	>6

Appendix V

Table 5: Molarity of the Concentrated Acid and Bases

Concentrated Reagents	Density	Molarity (M)	Normality (N)	Volume(ml) required to make 1000ml solution	
				1M	1N
Acetic acid 99.5% (formula weight: 60.05)	1.05	17.4	17.4	57.5	57.5
Ammonia (35%)	0.880	18.1	18.1	55.3	55.3
Ammonia (25%)	0.91	13.4	13.4	74.6	74.6
Hydrochloric acid 36%	1.18	11.65	11.65	85.8	85.8
Hydrochloric acid 32%	1.16	10.2	10.2	98.0	98.0
Hydrofluoric acid 40%	1.13	22.6	22.6	44.2	44.2
Nitric acid 70%	1.42	15.8	15.8	63.3	63.3
Perchloric Acid 60%	1.54	9.2	9.2	108.7	108.7
Perchloric Acid 70%	1.67	11.6	11.6	86.2	86.2
Orthophosphoric acid 85%	1.7	15.2	45.6	65.8	21.9
Sodium hydroxide 47%	1.5	17.6	17.6	56.7	56.7
Sulphuric acid 98%	1.84	18.4	36.8	54.3	27.2
Formic Acid 90%(HCOOH)	1.13	23.6	23.6	42.5	42.5
Ammonium Hydroxide	0.90	14.5	14.5	69	69
Sodium Hydroxide	1.53	19.4	19.4	51.5	51.5
Potassium Hydroxide	1.45	11.7	11.7	85.5	85.5

Appendix VI

Table . Atomic Weights of Selected Elements

Element	Symbol	Atomic weight	Element	Symbol	Atomic weight
Aluminium	A1	26.9815	Manganese	Mn	54.9380
Arsenic	As	74,9216	Mercury	Hg	200.59
Barium	Ba	137.34	Molybdenum	Mo	95.94
Boron	B	10.811	Nickel	Ni	58.71
Bromine	Br	79.904	Nitrogen	N	14.0067
Cadmium	Cd	112.40	Oxygen	O	15.9994
Calcium	Ca	40.08	Phosphorus	P	30.9738
Carbon	C	12.01115	Platinum	Pt	195.09
Chlorine	C1	35.453	Potassium	K	39.102
Chromium	Cr	51.996	Selenium	Se	78.96
Cobalt	Co	58.9332	Silicon	Si	28.086
Copper	Cu	63.546	Silver	Ag	107.868
Fluorine	F	18.9984	Sodium	Na	22.9898
Gold	Au	196.967	Strontium	Sr	87.62
Helium	He	4.0026	Sulfur	S	32.064
Hydrogen	H	1.00797	Thallium	T1	204.37
Iodine	I	126.9044	Tin	Sn	118.69
Iron	Fe	55.847	Tungsten	W	183.85
Lead	Pb	207.19	Uranium	U	238.03
Lithium	Li	6.939	Vanadium	V	50.942
Magnesium	Mg	24.312	Zinc	Zn	65.37

BIBLIOGRAPHY

Adam, G., and Duncan, H. (2001) Development of a sensitive and rapid method for the measurement of total microbial activity using fluorescein diacetate (FDA) in a range of soils. Soil Biology and Biochemistry, 33(7-8), pp. 943-951

American Society for Testing and Materials (ASTM) Standards (1966),p 824

AOAC. (1995). Official methods of analysis 16th Ed. Association of official analytical chemists. Washington DC, USA

AWWA, WEF, APHA, 1998, Standard Methods for the Examination of Water and Wastewater

Berger, K.C. and Truog, E (1939). Boron determination in soils and plants. Ind. Eng Chem. Anal. Ed., 11:540-545

Bouyoucos, G.J. (1962). Hydrometer method improved for making particle size analysis of soils. Agron J. 54: 464

Bray, R.H. and Kurtz, L.T. (1945). Development of total organic and available forms of phosphorus in soil. Soil Sci. 59, 39-45.

Brookes, P.C., Kragt, J.F., Powlson, D.S. and Jenkinson D.S. (1985). Chloroform fumigated and the release of soil nitrogen. The effect of fumigation and temperature. Soil Bio. Biochem., (17): 831-835.

Buurman, P.; Lagen, B. Van and Velthorst (1996). Manual for soil and water analysis Backhuys Publishers, Leiden pp. 5-6.

Casida LE, Klein DA, Santoro T (1964) Soil dehydrogenase activity. Soil Sci 98:371–376

Chapman, H.D. (1965) Cation Exchange Capacity. In: Black, C.A., Ed., Methods of Soil Analysis, American Society of Agronomy, Madison, 891-901.

Chesin, L and Yien, C.H. (1950). Turbidimetric determination of available sulphates. Proc. Soil Sci. Soc. Am., 14: 149-151.

Datta, N.P.; Khera, M.S. and Saini, T.R. (1962). A rapid colorimetric procedure for the determination of organic carbon in soils. J. Indian Soc. Soil Sci. 10: 67-74.

Esteban R.; Luis Balaguer ;Esteban Manrique , Rafael Rubio de Casas ,Rau´l Ochoa, Isabel Fleck ,Marta Pinto´-Marijuan, Isidre Casals,Domingo Morales, Marı´a Soledad Jime´nez, and Roberto Lorenzo, Unai Artetxe, Jose´ Marı´a Becerril, Jose´ Ignacio Garcı´a-Plazaola (2009). Alternative methods for sampling and preservation of photosynthetic pigments and tocopherols in plant material from remote locations, Photosynth Res (2009) 101:77–88, DOI 10.1007/s11120-009-9468-5

Flavel, T.C. and Murphy D.V. (2006) Carbon and Nitrogen Mineralization Rates after Application of Organic Amendments to Soil. Journal of Environmental Quality, 35, 183-193. http://dx.doi.org/10.2134/jeq2005.0022

Gaarder, T. and Graan, H. H. 1927. Investigation of the production of plankton in Oslo Fjord. Rapp Proc. Verb. Cons. Perm. Int. Explor Mer., 42: 1- 48.

G.J.S. Govett,(1961).Critical factors in the colorimetric determination of silica,Analytica Chimica Acta,25 (1),pp. 69-80, ISSN 0003-2670,https://doi.org/10.1016/S0003-2670(01)81519-1.

Greenberg, A.E.; Trussel, R.R. and Clesceri, L.S. (1985). Standard methods for the examination of water and waste water 16th Edition APHA, AWWA, WPCF Publication, U S A, pp. 1268.

Howell, J. 2005. Organic Matter: Key to Soil Management. http://www.hort.uconn.edu/ipm/veg/croptalk/croptalk1_4/ page8.html

Jackson, H.L. (1958). Soil Chemical Analysis, Prentice Hall of India Pvt. Ltd., New Delhi.

Jones, J.B. Jr. (1988). Soil Testing and Plant Analysis: Procedure and use. Tech. Bull. Food and Fertilizer Technology Center, Taipei City, Taiwan pp. 14.

Jhingran, V.G.; Natrajan, A.V.; Banerjea, S.M. and David, A (1988) Methodology on Reservoir Fisheries Investigations in India. Bulletin No. 12. Central Inland Capture Fisheries Research Institute, Barrackpore, West Bengal.

Kandeler,E. (2015). Chapter 7 - Physiological and Biochemical Methods for Studying Soil Biota and Their Functions, Editor(s): Eldor A. Paul, Soil Microbiology, Ecology and Biochemistry (Fourth Edition), Academic Press,PP. 187-222, ISBN 9780124159556, https://doi.org/10.1016/B978-0-12-415955-6.00007-4.

Kolthoff, I. M. and Sandell, E. B. (1952). Textbook of quantitative inorganic analysis, New York: The Macmillan Company, 759 P.

Keren, R. and Bingham, F.T. (1985). Boron in water, soil and plants. Advances in Soil Sci.: 414 – 415.

Knudsen, D., Peterson, G. A. and Pratt, P. F. (1982). Lithium, sodium and potassium. Pages 225–246 in A. L. Page et al., eds. Methods of soil analysis, Part 2. American Society of Agronomy, Madison. WI.

Lindsay, W.L. and Nervell, W.A. (1978). Development of DTP soil test for zinc, iron, manganese and copper. Soil Sci. Soc. Am. J. 42: 421-428.

Lorenzen, C.J. (1967) Determination of Chlorophyll and Pheopigments: Spectrophotometric Equations. Limnology and Oceanography, 12, 343-346.

http://dx.doi.org/10.4319/lo.1967.12.2.0343

Meyns S., Ill R., and Bibi B., (1994). Comparison of chlorophyll-a analysis by HPLC and spectrophotometry: Where do the differences come from? Arch. Hydrobiol. 132 (2),129-139

Novak JM, Lima I, Xing B, Gaskin JW, Steiner C, Das KC, Ahmedna MA, Rehrah D, WattsDW, Busscher WJ, Schomberg H. 2009. Characterization of designer biochar produced at different temperatures and their effects on a loamy sand. Annals of Environmental Science 3:195-206.

Olsen, S.R.; Cole, C.V.; Watanabe, F.S. and Dean, L.A. (1954). Estimation of available phosphorus in soils by extraction with sodium bicarbonate. Circ. V.S. Dep. Agric., 939.

Piper, C.S. (1966). Soil and plant analysis. Hans Publishers, Bombay, India pp 368.

Pickett, E.E. and Koirtyohann, S.R. 1969. Emission flame photometry—a new look at an old method. Analytical Chemistry 41 (14): 28A–42A. DOI: 10.1021/ac50159a003

Pillai, V.K., Boyd, C.E., 1985. A simple method for calculating liming rates for fish ponds. Aquaculture 46, 157–162

Prescot, J.A.; Taylor, J.K. and Marshall, J.J. (1934). Trans Ist Comm. Int Soc. Soil Sci. Versailles. Pp 143-153.

Shackley, S., Sohi, S., Ibarrola, R., Hammond, J., Mašek, O., Brownsort, P., Cross, A., Prendergast-Miller, M., Haszeldine, S., 2012. Biochar, Tool for Climate Change Mitigation and Soil Management. In: Meyers, R.A. (Ed.), Encyclopedia of Sustainability Science and Technology. Springer, New York, NY.

Schollenberger, C.J. and Simon, R.H. (1945). Determination of exchange capacity and exchangeable bases in soil ammonium acetate method. Soil Science, 133. 53-57.

Singh, Dhyan; Chhonkar, P.K. and Pandey, R.N. (1999). Soil, Plnat, Water Analysis – A methods manual. IARI, New Delhi. pp.6-11.

Sippola, J. and Ervco, R. (1987). Determination of boron in soils and plants. Anal. Lett. 8. 559-568.

Song, W. Guo, M. (2012). Quality variations of poultry litter biochar generated at different pyrolysis temperatures, Journal of Analytical and Applied Pyrolysis 94 , 138–145

Strickland, J. D. H., (1952). The preparation and properties of silicomolybdic acid; I. The properties of alpha silicomolybdic acid: American Chemical Society Journal, v. 74, p. 862-7.

Subbiah, B.V. and Asija, G.L. (1956). A rapid procedure for the determination of available Nitrogen in soils. Current Sci. 25. 259-260.

Tabatabai, M. and Bremner, J. (1969) Use of p-Nitrophenyl Phosphate for Assay of Soil Phosphatase Activity. Soil Biology and Biochemistry, 1, 301-307. http://dx.doi.org/10.1016/0038-0717(69)90012-1

Tabatabai, M. A. (1994). Soil Enzymes. In A. L. Page (Ed.), Methods of Soil Analysis: Part 2 Microbiological and Biochemical Properties (pp. 775-833). Madison, WI: The American Society of Agronomy. https://doi.org/10.2136/sssabookser5.2.c37

Takahashi, Masayuki, Ichimura, Shun-ei, (1970), Photosynthetic properties and growth of photosynthetic sulfur bacteria in lakes, Limnology and Oceanography, 15, doi: 10.4319/lo.1970.15.6.0929.

Tandon, H.L.S. (1993). Methods of analysis of soils, plants, waters and fertilizers, FDCO, New Delhi, pp. 143.

Van Heukelem., 2002. HPLC phytoplankton pigments: Sampling, laboratory methods, and quality assurance procedures. In: Mueller, J., and Fargion, G., (Eds.), Ocean Optics Protocols for Satellite Ocean Color Sensor, Revision 3, Volume 2, Chapter 16, NASA Technical Memorandum 2002-2004, pp.258-268.

Walkley, A. and Black, C.A. (1934). Estimation of soil organic carbon by the chromic acid liberation method. Soil Sci. 37: 29-38.

Watanabe, F. S. and Olsen, S.R. (1965). Test of ascorbic acid method for determining P in water and NaHCO3 extr of soil. Soil Sci. Am. Soc. Proc. 29 : 677-678.

Williams C.H. and Steinbergs, A. (1959). Soil sulphur fractions as chemical indices of available sulphur in some Australian Soils. Aust. J. Agric. Res. 10: 340-352.

Watson, C., Atkinson, D., Gosling, P., Jackson, L., Rayns, F., 2002. Managing soil fertility in organic farming systems. Soil Use Manag. 18, 239–247